MW00513296

BIOLOGICAL TRANSMUTATIONS

C.L. KERVRAN

 BEEKMAN PUBLISHERS, INC.

BIOLOGICAL TRANSMUTATIONS

C. LOUIS KERVRAN

Membre du Conseil d'Hygiène de la Seine

*Membre correspondant du Centre International
de Recherche Biologique de Genève*

Directeur de Conférences de l'Université de Paris

Ⓑ BEEKMAN PUBLISHERS, INC.
Woodstock, N.Y.

Published in the United States in 1998 by
Beekman Publishers, Inc.
PO Box 888 / 2626 Route 212
Woodstock, New York 12498
U.S.A.

All rights reserved. No part of this publication may
be reproduced, stored in a retrieval system, or
transmitted, in any form or by any means, electronic,
mechanical, photocopying, recording or otherwise,
without the prior written permission of the publisher.

Cover design and digital imaging
by Ra Z. Shakoor

First English edition, 1980, by Beekman Publishers, Inc.

© English translation by Crosby Lockwood, 1971

Originally published in French, 1966
by Le Courrier du Livre

English edition revised and edited by
HERBERT & ELIZABETH ROSENAUER

ISBN: 0 8464 0195 9
Library of Congress: 80-70745

CONTENTS

INTRODUCTORY NOTE

by Robert Waller, Editor, Soil Association Journal

There is an old saying that " the plant makes soil ". Fallow and rotations in the past did a great deal to maintain fertility without the use of fertilisers. The shifting cultivators of the early days of farming and in parts of Africa today would leave the soil to restore itself over a long fallow of up to twenty years or so. This restoration of fertility must be due to self-colonised plants (weeds) which return to the soil the nutrients that the sown crops had removed. Prof. Kervran says that daisies settle on lawns lacking in calcium and they " create " the calcium the soil requires. The plant, therefore, in this process of restoring fertility, puts into the soil minerals that it has not taken from it. We think of crops in our own monocultural farming as only taking from the soil, not feeding the soil. Prof. Kervran's experiments on biological mutations give an explanation of this phenomenon.

The Soil Association has joined with the publishers in presenting Prof. Kervran's book to an English public because they believe it is worthy of the attention in this country that it has received in France and Japan. They have not had the means of testing his theories for themselves. I myself know scientists whom I respect who do not agree with Kervran. I feel that only time and further research can decide. While at the Soil Association we have an open mind we can also see many interesting developments and implications of the theory. Empirically we know, as described above, that a skilful rotation of crops and sophisticated inter-cropping can do a great deal to keep up soil fertility—and soil structure. *Biological Transmutations* would provide a scientific explanation of this on which a great deal of further research into practical husbandry could be based. Our present economic pressures—or should one say temptations—make us depend too much on fertilisers to replace the effects of extractive farming. (And we do this on a very crude theory of plant nutrition.) The deterioration of soil structure already observed in this country may ultimately make us modify monocultural practices. A new scientific explanation of the relation between the plant and the fertility of the soil might then owe a great deal to Prof. Kervran. In France some agriculturalists are already working along these lines.

Whatever the ultimate fate of Prof. Kervran's convictions, we hope that they will provoke and stimulate agricultural research into thinking about some inexplicable and anomalous nutritional balance sheets which tend to be overlooked failing an explanation along orthodox lines.

PREFACE TO THE ENGLISH EDITION

This book was published in France at the end of 1966 and gave an adequate but not detailed description of the research into, and application of, the phenomenon of transmutation of elements by biological agents. In 1967 discussions took place for an English edition, and these were completed at the beginning of 1968.

I should like to thank Mr. and Mrs. H. Rosenauer and the Soil Association for arranging for the book's publication in English.

Many articles have been published in English-speaking countries about what I have termed, in an abbreviated form, " biological transmutations ", and this has been especially due to the disinterested and effective efforts of H. Rosenauer and H. Sabetay. They have published studies and discussions which have appeared in a number of periodicals; for example, in England, in the Journal of the Soil Association. Some have also appeared in journals in Canada, Australia and India, but until now a comprehensive detailed treatise in English has not been available.

I am sure that all who are interested in living creatures will find new perspectives as a result, and will look at nature with a clearer view. They will also realise that a system of farming which takes into account chemistry alone is a serious mistake which sooner or later must be paid for. They will also see that man and domestic animals are not adequately nourished on the diets formulated by chemists. Eventually such diets lead to deficiency diseases.

Professor R. Furon, of the Faculty of Sciences at the Sorbonne, reviewing this book for *Revue Générale des Sciences*, January-February, 1967, wrote: " This new work . . . reduces theories to an essential minimum, while discussing in detail observations and applications ". He admits that there is no other valid explanation than the transmutation of elements and adds: " We are definitely on a new path ". A Belgian professor in a discussion of this book in a teacher's journal, also remarked that transmutation of elements " is the only valid explanation " of the phenomena observed.

These phenomena have been noticed for a long time. They were first reported in Great Britain by Prout in 1822, and also by Lawes and Gilbert at Rothamsted in the second half of the nineteenth century, and have been utilised on the Soil Association's experimental farm at Haughley. A book has been published in Britain on these observations: *Continuous Creation* by Branfield, London, 1950. It reports certain observations hitherto incapable of explanation by classical chemistry, but to which I have been able to give an answer. This explanation has not been refuted by other experiments and, on the contrary, has not only been accepted by eminent scientists in many countries, but it has been taught in France at the most important Agricultural School for Engineer Agronomists (National Institute of Agronomy) and at the Faculty of Medicine in Paris where the author lectures.

It is evident that biological chemistry is mistaken in trying, exclusively, to apply chemical analyses to the study of living matter. When a molecule is taken away from a living cell it is impossible to study the cell's properties. The latter are dependent on the position of the molecule in a component and on the coupling of these components which, together, give rise to the many interactions characteristic of life. The desire to study a complex body tied by so much "feedback", while basing that study only on analyses of parts of the body, is so extraordinary that it is hard to understand why so many biological chemists do it. (Those who realise the error are very few.) Biological chemistry should be no more than a branch of biology, since biological processes are all based on synthesis. Biological chemistry is necessary, but is not alone sufficient. This is why the objections of biological chemists to the interpretation in this book of certain phenomena cannot be sustained; neither can the objections of physicists, who only take account of laws formulated by experiments on dead matter and on isolated atoms and molecules devoid of life. We are therefore in quite a new sphere and I trust the reader will be convinced of this when he has come to the end of the book.

C. LOUIS KERVRAN

WHAT ARE BIOLOGICAL TRANSMUTATIONS?

by Herbert Rosenauer

Classical science is confronted in agriculture with many enigmas which it believes can be explained by physico-chemistry; but this is presumptuous, and the great physicists are coming to realise that it involves over-hasty generalisations.

Louis de Broglie, the father of wave mechanics, wrote recently: "It is premature to suppose that we can reduce vital processes to the inadequate conceptions of physico-chemistry of the 19th and even the 20th century".

One could fill pages with quotations of this kind (and even with works on this theme) emanating from scientists from all over the world and showing that nature does not obey our physico-chemical laws; for nature does not work according to the simplified conditions that prevail in laboratories where the evidence is discovered from which we deduce the laws.

In chemistry we are always referred to a law of Lavoisier's formulated at the end of the 18th century. "Nothing is lost, nothing is created, everything is transformed". This is the credo of all the chemists. They are right: for in chemistry this is true. Where they go wrong is when they claim that nature follows their laws: that Life is nothing more than chemistry.

THE 19th CENTURY

As little as 30 years after the death of Lavoisier his law was challenged by the French chemist, Vauquelin. Sometime before 1822 when he retired, Vauquelin had shown that chicken which are only fed on grain, excrete more calcium than there is in the grain that they have ingested. This is impossible to explain by chemistry since the calcium appears to have created itself.

Soon after this, the Englishman, Prout, in 1822, made a systematic study of the variations in calcium in the egg of incubating chicken. He stated that the chick when it broke out of the egg contained four times more lime than there was in the egg: the weight in lime of the shell had not changed. Prout concluded that there had been endogenous formation of lime in the egg, a transmutation of an element which could become calcium (see the work *Transmutations à Faible Energie* p. 136 and following, by C. L. Kervran).

In 1849 in Germany, Vogel germinated some seeds of watercress and, despite the precautions taken, found more sulphur in the sproutings of the seeds than in the seeds themselves: he concluded from this that sulphur could not be a "simple body".

From 1856 to 1873 at Rothamsted, Lawes and Gilbert conducted some experiments showing that plants could abstract from the soil more magnesium than there was in the soil.

Von Herzeele, in Germany, from 1875-1881, made some systematic tests showing that, in germination, some elements increased, while others diminished: and he seems to have been the first scientist to devise experiments aimed to discover which element, by transmutation, was able to give another, as revealed by chemical analysis.

Thus the 19th century was marked by attempts—of which we have cited here only those which were undertaken the most scientifically— to confirm that there are in nature phenomena which are not explicable in chemical terms and apart from Vauquelin and Prout, the research was carried out with plants.

20th CENTURY

This research work was carried out mostly on plants because numerous observations made in agriculture could not be explained by chemistry.

This is why in all ages farmers have practised the fallow: when land is exhausted, it is given a rest . . . and whatever is lacking comes back on its own in a few years.

Pfeiffer, in Switzerland, and then in America, has cited many facts which do not tally with orthodox views. He has given these in his book *Fecondité de la Terre* (1935). For example, he has noted that the English lawn is good provided it contains calcium. But when the calcium is exhausted, then daisies appear. He has analysed the ashes of these daisies and they are rich in calcium. This led him to ask whence comes this calcium, since there isn't any in the soil? The question remained at that time unanswered.

There is a book entirely devoted to these anomalies—*Continuous Creation* by W. Branfield (London, 1950) and one could go on to cite other works.

In France, Freundler, lecturer at the Sorbonne (Faculté des Sciences de Paris) established in 1925 that iodine is created by algae (*laminaria*) and not—as has always been believed and is still taught today—by the algae fixing the iodine which is in the sea. A French geologist, A. de Cailleux, has, in addition, written that there is not a trace of iodine in the pre-Cambrian rocks which have remained unimpaired in their place: that being so, where does the iodine which is found in the earth later, come from? (Added to that, where do the immense masses of calcium come from found in the secondary epoch, if not, as in the primary, essentially from the siliceous rocks?)

One could go on indefinitely quoting such observations which are abnormal from the chemical point of view. Chemistry is only the exchange between atoms of peripheral electrons. Why does nature make no use of the rest of the atom, above all the nucleus, where almost all the matter is found?

No explanation of this was to be found. Freundler came nearest to it. (Von Herzeele had not been able to divine the mechanism, for at that time the nature of the atom was unknown.) But the works of Freundler came rather too soon. The neutron was not discovered until 1932, so that before then there was inevitably an incoherent element in the explanations.

THE WORKS OF C. LOUIS KERVRAN

From 1935 to 1955, in investigation into effects on man of the conditions in which he lives and works, C. L. Kervran was able to study some clear cases of intoxication caused by carbon monoxide when no one was able to detect CO in the air the men were breathing. It was not until 1955 that he spotted the close parallel between his cases and several other accidents that led to death in a surprisingly short time. This gave him the clue to the cause. From then on, for four years, he undertook systematic research which ended by confirming that nature is capable of transmuting elements. Not just any element, nor anyhow. Nature does not conform to our laws, she cannot do everything and all transmutations are not possible.

After that many examples came to Kervran's notice which remained in his subconscious, because he had not found adequate explanations and he systematically relied on varied research which led him to unravel the rules for the majority of the transmutations. He has made these studies the subject of four volumes. Volume one, *Transmutations Biologiques* (Lib. Medic. and Scientifiques, Maloine, Paris) was first published in 1952 and by 1955 it had already run through three editions. We hope that one book of his at least will soon be published in England if a publisher can be found. Two have already been published in Japanese in 1962 and 1963.

One cannot in a single article describe all the ramifications of this question of transmutations. We will limit ourselves to showing how it can provide an explanation of many observations made within different sciences working in the field of agriculture.

THE POTASSIUM BALANCE SHEET THAT DOESN'T ADD UP

According to Reinberg (*Le Potassium et La Vie*, P.U.F. Paris) in 1955 annual plants returned to the soil 1,500,000 tons of potash K_2O: but only 450,000 tons of K_2O had been brought to the soil in the form of artificial fertiliser plus about 300,000 tons of farmyard manure, and Reinberg concluded: " There appears to be an annual deficit of 750,000 tons . . . very partially filled in by the transformation of non-assimilable potassium into assimilable. Thus, the disappearance of a part of the non-assimilable potassium does not cover the deficiency, and the enigma remains. Reinberg has done no research into how this potassium could have been formed.

This mystery annoys the agronomists who naturally would like to make everything that happens in the soil conform to chemical laws. To resolve the problem once and for all a systematic programme of research was undertaken from 1961 to 1964 by the Station Agronomique de Dijon. The results were published in l'Academie d'Agriculture on the 12th January 1966. Here is a resumé:

Experiments in pots and in fields showed that plants return to the soil more potassium than there is exchangeable in the soil; the experiment had for its aim to prove that plants also return " fixed " potassium which only becomes gradually exchangeable as the weak acids of the soil (or some of the micro-organisms) work on it. In the outside plots (planted with cocksfoot—*dactylis glomerata*) the exchangeable K and the K insoluble by acid sodium tetraphenyl borate were measured; the fixed K is the difference between these two values. The measurement was made at the end of the experiment and the total K taken from the soil by the plant also measured. Potassic fertiliser had also been added. Thus it was expected to find in the plants in excess of added fertiliser, the difference between what remained in the soil as exchangeable K (against NH_4+) and as " fixed " K compared to what had been there at the beginning of the experiment.

Here are the figures for the two plots (expressed in K values in Kg per hectare)

Taken up by the plants	1,695	1,772.5
Lost by the soil (here the fertiliser is included)	1,534.5	1,629.5
	160.5	143

Therefore 10% more has been taken up by the plants than has been taken out of the soil. The experiment has convincingly shown that it is not only the exchangeable that diminishes but the fixed K also: there remains a deficit of about 150 Kg/ha, which confirms what Reinberg wrote: the reduction in fixed K only partially compensates for the deficit and there is a production of K which is inexplicable in chemical terms. The work of Professor Kervran has shown where this potassium comes from: it is created by the plant.

THE MAGNESIUM BALANCE SHEET THAT DOESN'T ADD UP

In 1947, Prince, Zimmerman and Bear, in New Jersey, on 20 different soils, observed that even if a plot appeared deficient (on chemical analysis) in magnesium, the quantity of magnesium increased in lucerne.

D. Bertrand, in his book *Magnesium and Life* (P.U.F. Paris) noticed that plants take from the soil 30 to 120 Kg/ha of magnesium. As we do not usually add magnesium fertiliser to the soil at this rate, the greater part of arable land would quickly become exhausted, which the experiment contradicts. But it is not only the fact that the exchangeable magnesium is increased by the addition of unassimilable magnesium becoming slowly available through the influence of the acid secretions of the plants or the micro-organisms which explains the phenomena, for all the exchangeable magnesium and the fixed magnesium would disappear quickly enough.

Prof. Kervran has also shown that man under conditions of intense heat, excretes more potassium than he receives, without exhausting his reserves, and he has shown the origin of this potassium which is created or manufactured by everything that lives.

THE NITROGEN BALANCE SHEET DOESN'T ADD UP

In his treatise on the microbiology of soils, Pochon (and de Barjac) of the Institut Pasteur, Paris, writes: In spite of the considerable number of works on the subject, the question of the balance of nitrogen in soils still raises a veritable enigma: whatever precautions are taken, when one adds up the difference between what is put in and what is taken out, there is always an important portion which cannot be recovered and which varies, according to the conditions, from 15% to 30%. He says besides: " The total balance of nitrogen is deceptive in the sense that it frequently shows a decline in relation to the evidence provided by the soils ".

At Rothamsted research work followed for 49 years has shown a loss of nitrogen of 51 t/ha (using NPK fertilisers). At Windsor (Connecticut) 10 years of continuous tobacco cultivation has given a loss of 17 t/ha with SO_4 $(NH_4)_2$ and of 18 t/ha with urea-cyanamide; in glass houses they have found a loss of 16% nitrogen non-recoverable.

None of these research workers has been able to see that if N_2 disappears, $C+0$ increases, a reversible reaction, according to the circumstances. Or rather, they have seen it, without seeing the meaning of it. Pochon writes: " in the fermentation of farmyard manure the ratio C/N which is 33:1 at the beginning falls to 13:15 at the end of the fermentation ". The work of Kervran explains how C can diminish while N increases. What accounts for this decline of the C/N ratio in this fermentation? In other cases we find a disappearance of N and an increase of the carbon compounds.

Research work since 1940 at the Haughley Experimental Farm in England (Soil Association) shows that where organic manures are used the reserve of nitrogen builds up much more than when artificial nitrogen is used: the micro-organisms, according to the theory of Kervran, have utilised the carbon and oxygen of the organic manures to create the nitrogen ($_6C+_8O=2._7N$).

MISCELLANEOUS

Earthworms are able to increase the calcium in a clay soil; they have glands that excrete $CaCO_3$. Wollny (1890), Puh (1941) Lunt and

Jacobsen (1944), Hopp and Clater (1948) have confirmed this. The actinomycetes (streptomyces especially) can also transmute silica into calcium. Here are some statistics taken from a publication of the French Academy of Sciences, 29th June 1964:

		SiO_2	CaO
Sandstone,	natural	63.0	1.40
do.	modified	35.8	17.34

Thus the lime has increased more than 12 times, whilst the silica has gone down (the other elements measured have not changed to a significant degree); rainwater running down to the bottom of a monument has been measured as well and no possible addition of calcium from an exterior source has appeared. If one calculates the difference between the values above, coming back to the atoms of silica and calcium and not their oxides, one obtains the compensatory formula $_{14}Si + _6C = _{20}Ca$*, a reaction brought about in the above case by the streptomyces which have been detected in 83 samples. This is the case in " stone sickness " i.e. disfigurement of stone by micro-organisms. In the soil it is the same process, and the explanation of the appearance of calcium in the secondary epoch is that it comes out of the silica of the primary.

CONCLUSION

The field of application of biological transmutations is immense and in this essay there was no possibility of doing more than giving a few examples in order to demonstrate that what is an enigma for chemistry becomes normal in the light of the laws of biological transmutations based upon the evidence put forward by C. L. Kervran as a result of innumerable laboratory experiments in making use of living matter, evidence which has confirmed and proved earlier findings. Nothing is served by denying these facts; agriculture bears witness to their existence. That classical nuclear physics cannot explain them is of no importance: its laws have been deduced *in vitro* having nothing in common with biology.

These experiments have been done with men, animals and plants and they reveal that enzymes—among them some of which have been identified—are specifically related to these transmutations, that these enzymes are found in the cells of animal organisms, or the higher plants or in micro-organisms: fungi, actinomycetes, bacteria, microscopic algae.

That is why it is indispensable to know about the works of C. L. Kervran in agricultural circles. These works have been the subject of lectures before many conferences of agricultural biologists in Paris,

* According to the system of symbolism used by Kervran.

Reprinted, with permission, from the Journal of the Soil Association.
Turin, Lausanne and Brussels, etc. and they are now taught by many lecturers in agriculture at colleges and institutes.

It should be emphasised that many control analyses were made in laboratories of the French Army and Navy.

INTRODUCTION

> "It is premature to want to assess vital processes according to the very insufficient physico-chemical concepts of the nineteenth or even the twentieth centuries."
>
> L. DE BROGLIE.

It would be presumptuous to believe that all living phenomena can only be explained by chemistry, even with the incorporation of certain physical laws as we know them today. Biochemistry and biophysics are only partial aspects of the living processes of plants and animals. A persistence in believing only in classical physical chemistry has led official science to many an impasse.

Early in 1959, after several years of systematic research, I decided to publicise my conviction, indeed my certainty, that there is a hitherto unknown property of matter which has been widely but unknowingly utilised. This new property, which I had demonstrated as a result of thousands of relevant analyses, gives living bodies the ability to transform not only molecules (which is within the field of chemistry) but atoms themselves. In fact, there is a transmutation of matter; a passage of one "simple body" to another, of one atom to another.

In spite of supporting evidence, there were sceptical persons whose turn of mind would not allow them to admit that someone else had found what they themselves had failed to observe—such dogmatists (and scientists are nevertheless fallible human beings) are fortunately a minority.

On the other hand, I received warm congratulations and encouragement from eminent scientists. They welcomed with evident satisfaction these firmly substantiated results, because in them were to be found the explanation for much which hitherto had remained obscure.

From the end of 1959, I received solid support from persons of true scientific spirit, who believed that such an overthrow of our accepted knowledge must not be impeded by the egoism and pride of persistent negationists. The latter trust only themselves; they refuse on principle to undertake the necessary experimental work, and continue their denials in order to avoid self-contradiction.

Some of the open-minded persons who were on the "reading committee" of a scientific journal agreed to ask me, at the end of 1959, for an article summarising my work. Thus, an initial step in the dissemination of these new ideas was made, and I am indebted to these prominent scientists who made it possible. The first step is always the hardest.[1]

Once this step had been taken, the general public soon became informed. Aimé Michel, in *Science et Vie* December 1960, devoted an article to my research; another appeared in January 1963, and in this way thousands of readers came to know of these new ideas. Publicity was also given on Belgian Television, December 1960, then in December 1961 on French Radiodiffusion (several wavelengths); then Radio Europe No 1, June 1961, and subsequent broadcasts in Belgium, Switzerland and Italy. The news of my discovery was carried in the industrial review *l'Usine Nouvelle* in 1961, in the review *Planète,* and in daily newspapers such as *Le Progrès de Lyon* in France and the *Corriere della Sera* in Italy. Thus hundreds of thousands of readers and millions of listeners became aware of " this genuine discovery ", an expression used by Professor Tanon in his preface to my first book, published in 1962. (Professor Tanon is at the Faculté de Médecine de Paris, vice-president 1963, president 1964 of the Académie de Médecine, president of the Conseil Supérieur d'Hygène Publique de France.) It is impossible to quote all the publications which, to my knowledge, have printed special articles on my work in France, Japan, Switzerland, Italy, Canada etc.

Following this publicity by radio and the press, I was asked to present a more detailed account; listening proved to be inadequate, as did several articles and talks in France and other countries (Belgium, Switzerland, Italy, etc.)

My book *Biological Transmutations* was published in October 1962 by Editions Maloine Paris. A second printing was necessary in May 1963, and another edition, with some modifications, appeared in May 1965. This had a preface by Professor Tanon, and a second preface by A. Besson, Inspector Général d'Hygiène and a member of l'Académie de Médecine. Therefore, in this first stage, I had the support of important sponsors.

In May 1963 I published a second book, *Natural Transmutations* as a supplement to the first. Jean Lombard, the geologist of world-wide reputation, contributed a preface and a second edition appeared in January 1966.

Then in 1964 *Low Energy Transmutations* embodied a synthesis of the first two books, and contained some supplementary material.

I compiled these books for a wide section of the public interested in scientific matters, not for the specialist. Some of the latter have asked me to consider writing a more advanced work; but this will have to be left until later. I have, however, been asked especially for an even more popular book, suitable for readers with a general secondary education, which would also be within the scope of those

[1] My thanks also to M. and Mme Galabert, the generous sponsors who took the initiative in arranging meetings for the presentation of my work to the public.

who have forgotten some of the scientific knowledge they once learnt. In addition, agriculturists from our large regional farm institutes and leaders of our farming industry have asked for a book which they could easily understand. A similar request has come from a number of medical men whose knowledge of chemistry, acquired as students, is already outdated; most of them were not told about nuclear physics, a subject not on the curriculum of many of the large medical schools even in 1965.

I hope that the following pages will meet the requests from these various sources.

The books mentioned above have not been translated into English. They are published in France with the following titles: *Transmutations Biologiques, Transmutations Naturelles* and *Transmutations à Faible Energie.*

Chapter 1

PRELIMINARY

" We are accustomed to having men jeer at
what they do not understand "

GOETHE.

(a) HISTORICAL

Lavoisier was a great French scientist at the close of the eighteenth century, and one of the creators of modern chemistry. His successors in the nineteenth century considered they could extend his laws, assuming them to be valid at all times and under all conditions.

From then on, what was taught became dogma and it would have been pretentious for anyone to have expressed divergent opinions. One general principle had been laid down—nothing is lost, nothing is created, everything is transformed. The atom, being the smallest particle of matter, constituted a constant of Nature. For example, one could not create an atom of calcium; it could neither be lost nor disappear; if it had ceased to be part of one molecule (an assembly of two or more atoms) it could be found in another molecule—there had been a transformation of molecules, a chemical reaction. This was a *credo*, undisputed by applied science in all countries in the nineteenth century.

Millions of experiments performed throughout the world ensured that whoever doubted this established principle would bring ridicule upon themselves. In fact " Science " shrugged its shoulders at such sceptics in disdain and said nothing, considering it useless to argue with a pretentious ignoramus who might claim that there were cases where Lavoisier's laws had not been verified. In this way certain experiments of great interest were ignored and unnoted.

It was not until the twentieth century that this law of invariability of matter, undisputed for 100 years, received its first officially recognised breach. This was through the discovery of natural radioactivity which showed that certain bodies could change into something different. The alchemists of the middle ages who affirmed the existence of transmutation had been jeered at in the eighteenth century, especially so in the nineteenth, and even so today.

The unstable radium atom (radioactive) is finally transmuted into a stable atom of lead (non-radioactive). Several elements exhibit this property; these elements are rare because they rapidly disappear. One half of the radium disappears in 1620 years; half of the remainder,

1

constituting one quarter of the original mass, will disappear in another 1620 years, and so on. Some elements had an even shorter life, and have completely vanished from the earth's crust. The absence of these substances is one reason why this phenomenon had not been observed earlier; also, methods for identification of radioactive bodies were only developed towards the end of the nineteenth century. (We note that certain radioactive substances are being continually manufactured by disintegration of very long-lived substances. Radium originates from uranium-238, which has a half-life of 4.5 billion years. After giving off lighter and still lighter radioactive elements, radium eventually becomes a stable lead isotope.)

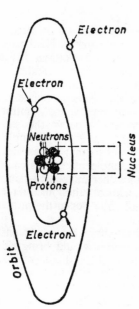

Fig. 1. Diagram of lithium atom. $\frac{7}{3}$ Li; 3 electrons, 3 protons, 4 neutrons. (There is also an isotope with 6 nuclear particles: 3 protons and 3 neutrons).

NOTE: All atoms except hydrogen have two electrons in the K-orbit, that nearest to the nucleus; see also Fig. 2.

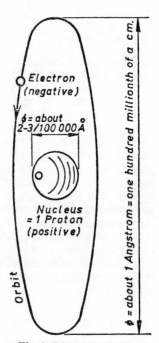

Fig. 2. Diagram of hydrogen, the simplest atom. (1 proton and 1 electron).

It should be noted that what chemistry calls an " element " is called a " body " in current terminology, and this is the old scientific name. Oxygen, sulphur and copper are elements. An element is composed of atoms, each one possessing the same number of protons. These are heavy particles with a positive electric charge and, together with neutrons, they make up the atomic nucleus; a similar number of lighter particles, negatively charged, revolve in separate orbits around the nucleus. An atom is electrically neutral since it has the same number of protons as electrons (Fig. 1).

The elements were formerly called " simple bodies ". They could not be made by man and have existed in Nature since the creation of our planet. They could not be divided or broken up and represented the simplest possible unit, hence the name " simple body ". It was only possible to transfer them from one molecule (or compound body) to another molecule. It is to these simple bodies that Lavoisier's Law of Conservation of Matter was applied.

It had been established that there must be ninety-two natural elements. This was arrived at by deductions made from certain similarities in properties observed by Mendeléev in the mid-nineteenth century. He arranged the then known elements into a table (of two entries). Some positions in the table were left blank, but were gradually filled as new elements were discovered with properties which corresponded to those allowed for in the table. If some positions still remained unfilled, it was partly because the missing very rare elements had not yet been isolated, and because some of them no longer existed on earth. These were the radioactive elements. They had been transmuted long ago into other elements, but this was not realised until the twentieth century when it became known how to reproduce these vanished radioactive elements artificially by means of nuclear physics. (Alternatively, these elements perhaps never existed in a natural state and can only be reproduced in the laboratory by particle acceleration techniques.)

However, this breach in Lavoisier's law only concerns radioactive bodies, which are an aspect of nuclear physics.

In chemistry it is still taught that nothing is lost, nothing is created, and that in any *chemical* reaction it cannot be otherwise. This I freely accept.

(b) CHEMISTRY AND ITS LIMITATIONS

The mistake has been the contention that all reactions occurring within living organisms are chemical reactions, and that life could and should be interpreted in chemical terms—hence the sciences of biochemistry and biological chemistry. Many living processes certainly take place by means of chemical reactions, but it is a fallacy to believe that there are *only* chemical reactions and that every observation can be explained by them.

The subject of this book is a property of matter which had not been noticed, a property which pertains neither to chemistry nor nuclear physics in its present state. The difference between the new science studied here and the two classical sciences will be duly explained, and the laws of chemistry will not be questioned. Far too

3

many chemists and biochemists have sought to apply, at all costs and without proof, chemical laws in a field where chemistry is not entirely valid or else is only an end phase. Such a phase follows a hitherto unobserved phenomenon. It is with this particular phenomenon that this work is concerned.

Chemistry is known to everyone; some of its rudiments are taught in primary school, and it will be easy for us, therefore, to suggest comparisons with chemistry. On the other hand, nuclear physics, being a new subject still undefined in many aspects, is less known. I shall merely state here that it is not applicable to the phenomenon I shall describe, in which a completely different science is involved. The reader wishing for more detailed knowledge is referred to my book *Low Energy Transmutations*. In an addendum to this book is a chapter explaining why the current laws of nuclear physics are not valid for the new science outlined in the following pages.

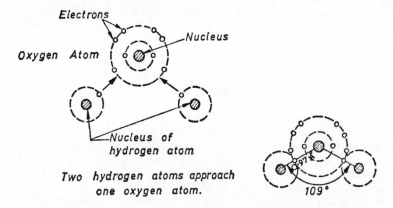

Fig. 3. Formation of the water molecule H_2O. The electron of each hydrogen atom enters the outermost orbit of the oxygen atom. which then has eight electrons in its external layer.

NOTE: One of the hydrogen nuclei, in the form of a positively-charged free proton can be separated under various conditions; it is called the H^+ ion; the remainder has one more electron than protons it is therefore negatively charged, this ion is written $OH-$. Together they are an electrolyte.

(c) BRIEF EXPLANATION OF SYMBOLS USED IN CHEMISTRY

Although everyone may know a certain amount of chemistry I will avoid the use of chemical formulas as much as possible. And in order to avoid writing the full names of elements each time, they will be represented by international symbols. The names of elements are often the same, or nearly so, in the majority of languages, and they are accordingly represented by their initial letter as a capital: O = oxygen; H = hydrogen; P = phosphorus; S = sulphur, etc. The formula of the water molecule is H_2O, signifying two atoms of hydrogen combined with one atom of oxygen (Fig. 3).

Where there are several elements with the same initial letter, they are distinguished by adding to the capital letter a small letter taken from the name of the element. Thus: B = boron, Be = beryllium, Br = bromine; F = fluorine, Fe = iron (French, fer); C = carbon, Ca = calcium, Cu = copper (French, cuivre), Cl = chlorine; Mg = magnesium, Mn = manganese; A = argon, Ag = silver (French, argent), Al = aluminium, As = arsenic; S = sulphur, Si = silicon, etc. The symbol for nitrogen is N; that for sodium is Na (derived from natrium). For potassium the symbol is K, which comes from the Arabic *al Khali,* a strong base, and is exemplified by the expression volatile alkali, meaning a volatile base, e.g. ammonia in current terminology. (Ammonia is not a simple body, but a chemical combination of nitrogen and hydrogen with the formula NH_3. This is a gas, soluble in water, the solution being called ammonia water.) In medical terminology there are also instances of two words having the same meaning, e.g. potassemia or kalemia with their compound forms prefixed by hypo- or hyper-. We can also mention gold, symbols Au from the Latin *aurum,* and mercury, Hg from hydrargyrum, the root of which exists in the name of the illness due to mercury, hydrargyrism.

(d) SYMBOLS USED IN NUCLEAR PHYSICS

When one speaks of the nucleus of an atom, the chemical symbol for an atom is not always adequate, and one then indicates the total number of particles in the nucleus (protons + neutrons which together constitute the nucleons or nuclear particles). This is because natural elements are varying mixtures of atoms; all atoms of the same element have the same number of protons and electrons since a normal atom is electrically neutral. It is this fact which characterises an element; but it has been shown that the number of neutrons in a nucleus can vary. The neutron is so called since it is (electrically) neutral, and it is almost the same size as a proton. Moreover, a neutron can be linked to a proton which might have a negative charge, thus neutralising the positive charge. This happens because a neutron is a little heavier than a proton. The foregoing is a simplified explanation. (The positive electron is also called a positron.)

Atoms with the same number of protons but a different number of neutrons are called isotopes. For example, the stable (non-radioactive) isotopes of magnesium are written: ^{24}Mg, ^{25}Mg, ^{26}Mg, which signifies that there are 24, 25, or 26 nucleons in the nuclei. Since all magnesium atoms have 12 protons, the number of neutrons is arrived at by subtraction. In the same way there are ^{10}B and ^{11}B for boron; ^{12}C and ^{13}C for carbon; ^{28}Si, ^{29}Si and ^{30}Si for silicon, etc. The diagram (Fig. 4) shows isotopes 6 and 7 of lithium.

5

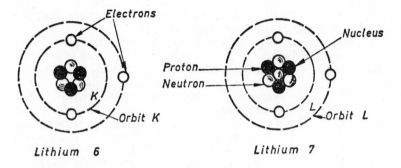

Lithium 6   Lithium 7

Fig. 4. Diagram of the two stable isotypes of lithium.

NOTE: The electron orbits are also termed "layers", the nucleus spins on its axis, and the electrons rotate in layers surrounding the nucleus. The diameter of the K-orbit decreases with an increase in the number Z (atomic number) of protons.

Text books often give notations such as

$$^{16}_{8}O, \quad ^{17}_{8}O, \quad ^{18}_{8}O,$$

representing the three stable isotopes of oxygen. In each case the subscript figure 8 signifies the presence of 8 protons (and 8 electrons). If there are 16 nucleons and 8 protons, then the number of neutrons 16–8 = 8. For isotope 17, the number of neutrons is 17–8 = 9, and for isotope 18 is 18–8 = 10. Isotope 16 is the most common (99.76 per cent).

(e) CHEMICAL COMBINATIONS

The phenomenon of biological transmutation involves the atomic nuclei. However, the most common and important of these transmutations concerns the first 20 elements especially, and to a lesser extent the next 10 out of a total of 92. (There are also a few artificial elements, called "transuranic elements" of an atomic number higher than 92, uranium.) Elements are placed in the order of their respective number of protons (and electrons), the first being hydrogen, H, which has only one proton; then helium, He, with two; then lithium, Li, with three. Here, with the appropriate symbols, are the first 20 elements. A few of them are not concerned in biological systems, or at least not so far as is known. These are: helium, beryllium, neon (Ne), and argon (A). The most common isotope is given.

6

1							4
H							He
1							2

7	9	11	12	14	16	19	20
Li	Be	B	C	N	O	F	Ne
3	4	5	6	7	8	9	10

23	24	27	28	31	32	35	40
Na	Mg	Al	Si	P	S	Cl	A
11	12	13	14	15	16	17	18

39	40
K	Ca
19	20

The right-hand column gives the rare gases (helium, neon, argon) which take no part in any of the reactions we shall deal with. These elements on the right play no part in chemical combination—they are said to be " saturated " since their outermost atomic layer contains 8 electrons, except for helium which has only one layer and that is saturated with 2 electrons. This is a stable position—except for some very rare exceptions—for the heaviest gases which may lose an electron through being caught by fluorine which has 7 electrons and thus gets 8, in a chemical combination with krypton, xenon, but excepting this case the rare gases always have 8 peripheric electrons and this number cannot vary.

The elements in the left-hand column have one electron in their outer layer; (fig. 5), those in the second column from the left have 2; those in the third have 3, etc. The elements of the 7th column have 7 electrons; as the natural tendency is to gain stability with 8 electrons (we do not know why, it just happens to be so) when the atoms of the elements of the 7th column are situated where they can " adhere " to an atom of another element, they " bind " themselves to it so that the molecule thus formed has a peripheral layer of 8 electrons. All the elements of the 7th column can take only one electron and are termed monovalent.

The elements in column 1, have only a single peripheral electron in their orbit; this electron can easily be detached and is the only one which can be released. All elements in column 1 are therefore monovalent. Those in column 2 can give up 2 peripheral electrons, they are divalent (or bivalent); those in column 6 are also divalent since they can take 2 electrons to complete the 8 of their peripheral layer. It is the same thing for the trivalents in columns 3 and 5; there are donors and receivers. It will be observed that there can be molecules with H (monovalent) and F (monovalent) giving HF (hydrofluoric acid); Na and Cl giving NaCl (sodium chloride or common salt or sea salt) etc. However, for water, since oxygen is divalent, 2 hydrogen

7

atoms are required to supply the necessary electrons, thus its chemical formula is H_2O. Similarly we have H_2S = hydrogen sulphide; KCl = potassium chloride; CaO = calcium oxide or quicklime; NH_3 = ammonia; CO_2 = carbon dioxide. In the latter, one atom of tetravalent carbon combines with two atoms of divalent oxygen (here it should be noted that the elements in column 4, being at the centre, are in a somewhat different situation; they can give up 1, 2, 3 or 4 electrons, or can take on the same numbers. This is why carbon has many possible bonds. Certain elements of columns 3 and 5 can also be donors or receivers of electrons; that is, they act either as reducing or oxidising agents.)

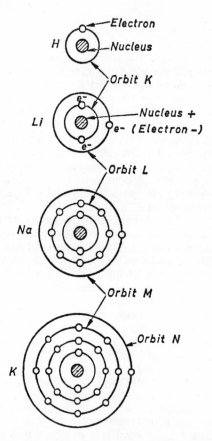

Fig. 5. Diagram of the first four elements of the first column of Mendeléev's table. (All are monovalent, with only one electron in the orbit, the outermost layer.)

8

This brief revision will perhaps be useful to those who have lost sight of certain elementary ideas of chemistry. It is quite rudimentary, only the facts necessary for an understanding of this work are given, and we will avoid, as far as possible, any reference to them. The intention is not to consider theoretical explanations, but essentially facts which cannot be explained by chemical reactions, and to this end a comparison with chemistry will on occasion be useful.

At this point it should be said that the concept of pH as a convenient means of defining alkalinity or acidity has been found to be an over-simplification and, at times, useless in certain aspects of biology.

In effect, pH is used to evaluate the acidity or alkalinity of aqueous solutions, but most organic liquids are not solutions in pure water; there are organic acids and bases. A strong mineral acid (in water), such as hydrochloric acid, is only partially broken down (dissociated or ionised) by formic acid; a weak acid such as monochloracetic acid behaves as a strong acid in a more alkaline solvent such as ammonia.

Therefore the concept of acidity-alkalinity has been entirely upset, and studies have been undertaken with the purpose of characterising acids and alkalies in relation to the medium in which they exist. As defined by Bronsted, an acid is any molecular structure capable of donating a proton, and an alkali is a structure which can receive or bind it. (This theory of Bronsted has been supported by Lowry.)

It is now accepted that hydronium (H_3O^+) is an acid since it can give up a proton to the basic hydroxyl OH^- which then becomes H_2O. H_2O is thus either an acid or a base, depending upon the medium, because in giving up one proton OH^- remains, and in taking a proton up there is H_3O^+. (Water is therefore amphiprotic.)

This book is intended for readers whose chemical knowledge is perhaps not up-to-date and I will keep to the old simplified ideas. It must not be forgotten, however, that reality is complex, and not always accessible to our present state of knowledge. This is so even in apparently simple cases. For instance, no one knows why in electrolysis hydrogen and oxygen appear at the electrodes in a gaseous form when their ions are liquid, and no evaporation is involved.

(f) AN EXAMPLE OF THE SHORTCOMINGS OF CHEMISTRY —ACTIVATED WATER

To show that chemical analysis is inadequate in determining the biological properties of a substance, let us have a brief look at the case of activated water.

All specialists in hydromineral treatments know that it is dangerous to consume excessive quantities of the gushing water from the spring of a spa. One must accustom oneself, progressively, beginning, for example, with a quarter of a glass at each session in order to avoid a violent shock to the system. On the other hand, the same water bottled and drunk several weeks later can be taken in quantity without having any other ill-effect than tap water. This fact remains unexplained; radioactivity has been suggested as responsible, but this has not been proved.

It cannot be denied that water flowing underground acquires special properties. Yves Rocard, director of the physics laboratory at the Ecole Normale Supérieure in Paris, has shown that water sets up an electromagnetic field which can be detected by a very sensitive magnetometer. It is this property which permits diviners to detect water running between particles of rock in the ground. Water, therefore, can be said to acquire electrical properties.

As is known, the electrical charges of a water molecule are asymmetric, they act as a dipole and behave within the magnetic field as would a small particle of iron filing.

However, " activated " water is not perhaps water with arranged molecules. We know nothing of the physical disposition of molecules in this activated water. It can be produced empirically by various methods; for example, by electromagnetic fields of very low frequencies (around 10 periods/sec. or 10 Hz), but sometimes fields of higher frequencies are utilised around 10kHz (or 3 to 4 kHz), but super-imposed on frequencies which differ very little in order to obtain a low frequency impulse. There are also other means.

The effects of this water have been studied in Italy, especially, by Professor G. Piccardi, of the Institute of Physical Chemistry in Florence, following observations made in 1935 by Beccari. Activated water, when heated, removes scale inside boilers; the adhering cal-careous substance (the scale) is insoluble in ordinary water, but this same water, activated, precipitates this calcareous substance in a colloidal form. *Chara foetida*, an algae having a calcareous mem-brane, does not change its form in ordinary water where it lives, but in the same water, activated, it loses its coating by precipitation.

It is evident that activated water can have important effects on the organism, as living cells are mainly composed of colloids.

The example cited above is only one special instance of the be-haviour of water. There is no substance which has been the object of so much study and which is still so imperfectly known. The effects of water are both complex and disconcerting; its properties may vary according to a number of factors which so far have not been clarified in spite of the studies of G. Piccardi and his pupils, and others such as F. Valfré, Maletto, etc. in Italy, Belgium, Switzerland and Ger-many. I refer the reader to that excellent book by G. Piccardi (in English) *The Chemical Basis of Medical Climatology*. Its author questions whether life could be possible without these variable proper-ties of water. The intrinsic energy of water is quite variable, at times it has a negative entropy and at other times a positive entropy. It changes its properties between 30° and 40°C in such a manner that around an average temperature of 35°C certain of its physico-chemical and biological properties are " stimulated ".

Accordingly, it may be asked if it is not when water shows a negative entropy that it furnishes the energy necessary for biological transmutations. But in the field of biological energy, in spite of numerous researches into the phenomenon, we still do not understand matters clearly. In any event, the behaviour of water is a good illustration of the fact that a chemical analysis is incapable of revealing the total constituents of a substance with certainty—even of the commonest in Nature, water. The formula H_2O could only be a simplification, perhaps admissible in the state of steam, but certainly not in the liquid state where there are ionised molecules. Water has been proved to contain $(H_3O)^+$, though in a small quantity; it also appears to contain $(H_5O_2)^+$, $(H_7O_3)^+$, etc. The physics of water is extremely complex and it would be impossible for me to give a brief account of it in this chapter. One should even say " waters ", rather than speak of it in the singular, and there exist waters of varying molecular structures on which depend the divers forms of ice crystals.

Let us bear in mind, especially, that chemical analysis, alone, is inadequate in biology, and consider the case of heavy water (where a neutron is present in the hydrogen nucleus). If the proportion of heavy water in ordinary water exceeds 50%, all life becomes impossible except for certain bacteria of the genus *Pseudomonas*. In some organo-metallic solutions, such as Grignard's reagent (ethyl-magnesium-bromide), the magnesium-25 (stable) is exchangeable; but its isotope 28 (radioactive) is not. Chemical properties can therefore vary with the isotopes just as the biological properties do. (Care is needed in the interpretation of results obtained with radioactive tracers, one can never be sure *a priori* that stable isotopes always behave in the same way.) We must avoid referring everything in biology to the laws of chemistry or even to those of physico-chemistry, since there are many unknown factors involved in the reactions activated by enzymes. We do not know the answers to many of the problems set by biological molecules. One is reminded of the work by Pasteur on polarised light, where he showed that polarisation distinguishes living from dead matter, even if the chemical composition is the same[1].

In biology it would be quite presumptuous to believe that we know much about the behaviour of the atom. The door which I am opening only reveals a part, and it is not an unknown aspect of chemistry that appears, but a new science involving the nuclei of atoms which might even be called " infra-atomic ".

The behaviour of colloids in " activated " water has been the subject of research throughout the world and attention should be drawn to the investigations in France made by A. Bérard and Dr. J. de Larebeyrette. They have shown that artificial media containing an organo-metallic compound can constitute patterns of sensitive biological tissues which are simpler, and therefore more easily studied,

[1] Similarly in photophoresis; there is an helicoidal rotation to the right for living proteins, and to the left for dead or synthetic proteins. L. C. Vincent has said " the chemical industry cannot make right-hand proteins ". Also, chlorophyll, in photosynthesis, turns spirally to the right.

11

than living tissues. Such media replace " a complicated natural phenomenon, impossible to reproduce, by one which is simpler and repro-
ducible, which can be made to evolve in accordance with certain
known parameters ". Here we approach what has been termed a new
aspect of " physico-chemical biology ".

Of course, one only needs to realise "conditions which imitate
more or less the essential conditions (of complex biological processes)
where different factors evolve which cannot be expressed by equations." But the investigations by Bérard and de Larebeyrette have
shown that methods of stimulation other than electric (heat and
light) can produce measurable reactions in these organo-metallic
colloids. It is also of interest to note that an activated liquid, depending on its composition and the nature of the metal of the electrodes,
may only lose its stimulation after 200 hours (whether activated for a
few seconds or a few minutes) and gives a discharge (deactivation)
with peaks of up to 500 mV.

Moreover, it is a curious fact that the restored energy is greater
than that put in (shown by integration of surfaces indicated by the
curves registered on an oscillograph) so that the negative entropy
of living tissues reappears. Bérard and de Larebeyrette believe that
a " liberation " of part of the potential energy contained in the
sensitive media is involved. The exact form of liberation remains to be
studied.

All of which shows that the phenomena studied by Piccardi and
others relative to activated water and used by Violet for his " dynam-
ised " electro-vibrated water, etc. are the subject of more detailed
investigations at a wider range.

It is possible to think that the origin of biological transmutations
could be this " liberation " of energy, under particular conditions, in
the colloidal medium of living cells when coenzyme metal is present
(statement made by Dr. de Larebeyrette at an experimental demonstration of his work at the Cercle de Physique A. Dufour, in May
1965).

(g) EXAMPLES TO SHOW HOW PHYSICS CANNOT BE APPLIED TO BIOLOGY

(1) It is a mistake to think that biology can ever be entirely
explained in terms of chemistry; it is likewise a mistake to claim that
physical laws are necessarily the same for living as for inanimate
matter.

In 1936, I published the first results of my researches into the
prevention of accidents caused by electricity. I showed that the human
body does not follow Ohm's Law, and that the application of this
law to man had led to serious mistakes. The body's resistance to an
electric current varies with a diversity of factors, such as the strength
or voltage of the current, and a strict application of Ohm's Law to
calculations for the human body is an absurdity which could have

fatal consequences. The human body (and that of animals) in no way behaves as a metal resistance, so that this simple law of physics is invalid in a biological context. With the aid of many experiments, the results of which were published from 1936 to 1940, I went very deeply into this question and my results were later confirmed by others. The importance of the research has now been recognised throughout the world (international conferences of engineers accepted it in 1938-39) and the subject is now taught in schools of medicine (see, for example: *Précis de Médecine du Travail*, by Professor Simonin, Maloine, Paris).

(2) We have seen that the principles of nuclear physics are certainly not all valid in biology. Thus the energy of an irradiation (α, β, γ, neutrons) measured by instruments used in physics takes no account of biological effects, so much so that, in biology, a unit of measurement different from the roentgen has had to be adopted: the "rem". This biological unit is somewhat imprecise, since no instrument can measure it directly. The effect produced by the roentgens is registered on the instrument, but a roentgen of radium, or cobalt 60, or thallium or X-ray is not of an equivalent value in biology. (In biological work the term "rad" is often used. This unit is a physical one, but is utilised comparatively since physiological effects vary from one tissue to another.)

(3) Similar remarks apply when interpreting the measurements of instruments such as light meters and sound meters. The human eye and ear do not have the same reactions as these instruments.

(4) The reluctance of some physicists to admit that not all their laws are applicable to biology, comes mainly from those University professors of physics who have lost sight of original principles on which the laws were founded. They declare that a negative entropy, the force which in biology would build up matter, is an impossibility. For them there is only positive entropy, in accordance with the second principle of thermodynamics of Carnot-Clausius regarding the breakdown of energy. They maintain that energy always breaks down; there can be no "negative energy" in the opposite sense of the law of degradation. This would imply an endogenous production of energy within the organism to a level which, in their view, tends to decline naturally towards the breakdown and death of matter.

These physicists do not deny the presence of an energy which maintains life, but for them this energy comes from what the organism takes in from its surrounding outside medium. They do not realise the great flaw in their argument! If this energy comes from outside, it is a mistake to represent it as justification for applying Carnot-Clausius' principle to biology. This principle has never been applicable in such cases. It was formulated only for closed systems having no exchange with the outside; therefore the concepts of entropy and negentropy (negative entropy) have no significance when there is an exchange with the outside medium. Unfortunately, this terminology has been too generally applied to a new technique, that of cybernetics. This has caused confusion, as the words do not have the same meaning. One starts from a similarity and forgets that it is only a similarity, and that the cardinal point of Carnot's principle is that it applies to closed systems. Many of the current confusions in science arise from differences in the correct use of words.

(5) Who is to say in which present-day branch of physics " mental energy ", the strength of will or character, should be placed? One can associate memory with information, and negative entropy with cybernetics (or should it be chemistry?), but nothing tells us if intelligence itself will not some day be expressed by a physical or chemical law. We should not try to explain everything in terms of our present knowledge, but remember instead the advice of de Broglie in the quotation heading the introduction to this book.

Chapter 2

EXAMPLES OF FORTUITOUS OBSERVATIONS

(1) IN MY YOUTH

Fortuitous observations are common enough; Branfield has made a book out of them: *Continuous Creation,* so has Czapeck, Hauschka, etc.

Personally, I made many such fortuitous observations during my youth. Some of them are given below.

(a) *On Hens*

My parents kept some hens in a shed with free access to a yard. We lived in central Brittany, where my father was a civil servant. The district was one of schist and granite, devoid of limestone. Limestone was never given to the chicken, but every day they produced eggs with a calcareous shell. I never thought of asking where the calcium of the egg came from (nor the calcium in the bones of the birds). But I was intrigued by an observation I had made. When the laying hens were loose in the yard, they pecked incessantly at the flakes of mica which dotted the ground. (Mica, together with quartz and feldspar, is a constituent of granite. They are all compounds of silica. That was all I knew at the time, when I was at primary school.) I noticed this evident selection of mica by the hens when the sun was shining after a shower. Well washed by the rain, the hundreds of flakes visible in each square metre looked like miniature mirrors, and the pecking of the hens was easy to follow. No one could tell me why the birds pecked the mica and not the grains of sand. I watched my mother opening the gizzard after a fowl had been killed and saw small stones and sand inside, but never mica. Where had the mica gone? This made an impression on me, and like everything which remains a mystery, stayed in my subconscious mind. Being a child, I liked solid logical explanations—the reason why.

(b) *On Incandescent Stoves*

In the district where I lived, our classroom was heated by a rustic cast iron stove without any fire-proof lining. There was a damper on the stove pipe for regulating the draught; alternatively one opened or closed the firebox. The fuel was split logs of old oak. After the wood was well alight, the stove very soon " roared " and glowed red-hot. Thereupon everyone complained of a headache, and a prefect had to push in the damper or close the firebox. The teacher told us the headaches were caused by the carbon monoxide fumes from the

15

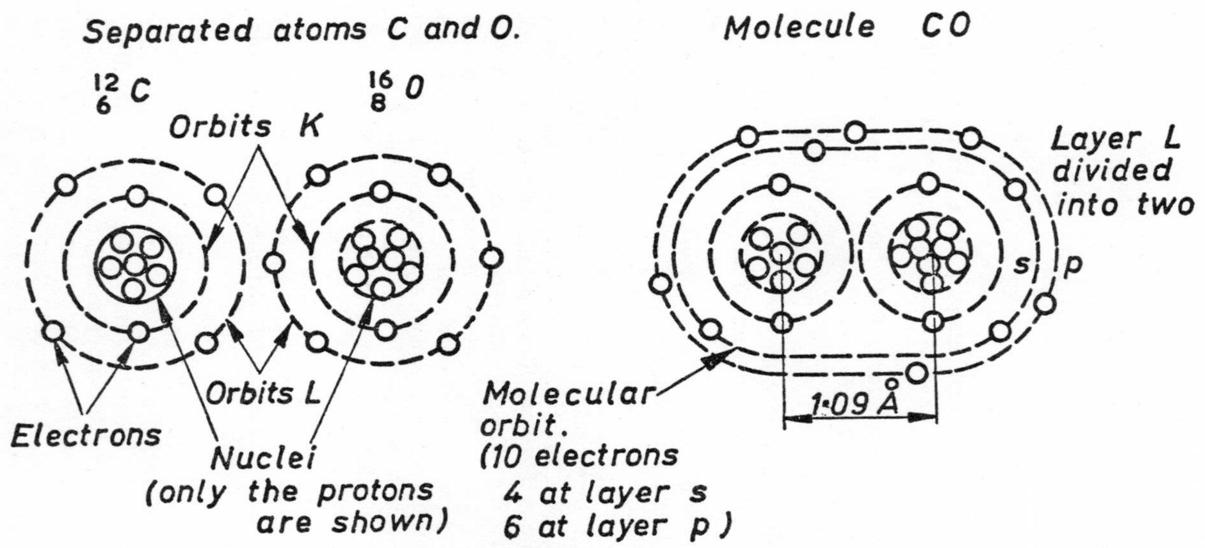

Separated atoms C and O.   Molecule CO

$^{12}_{6}C$ $^{16}_{8}O$

Orbits K

Orbits L

Nuclei

(only the protons are shown)

Electrons

Molecular orbit. (10 electrons)

4 at layer s
6 at layer p

Layer L divided into two

s   p

1·09 Å

Fig. 6. Another molecular representation.

The linking of the peripheral electrons in a chemical reaction is an easy way of explaining valencies; actually it is more complex. The two nuclei of the atoms C and O approach closely, each keeping its electron layer K, while the other electrons (10 for the CO molecule) arrange themselves in a molecule orbit around the "core" composed of the two nuclei.

In a reaction of biological transmutation, the two nuclei approach each other even closer, the distance between them is certainly 100 times shorter and probably less than 1000 times (here it would become Silica $^{28}_{14}$Si); an electron is ejected from each K orbit, and these enter rearranged L and M layers.

16

red-hot, cast-iron stove. There was no reason to doubt this, since one experienced the symptoms of carbon monoxide poisoning. (Fig. 6).

During science lessons (lessons of " things of the past ") we were told that it was dangerous to have slow combustion stoves in bedrooms, because carbon monoxide is formed when combustion is slow, whereas when it is rapid, carbon dioxide is formed, which is non-toxic.

All the answers to the questions which I later asked my teachers remained so unconvincing that here again was something that " stayed in my mind ".

(2) DISCUSSION ON CARBON MONOXIDE PRODUCTION

Later we shall see the explanation, but first we will cite some observations which once appeared mystifying, because they could never be explained. In such cases, one says modestly that there are some things that escape us, our ignorance is still great, and as no one can give a valid explanation it is useless to consider it further. All the more reason to do so since scientists have not found the answer. Or too often an explanation is given which is no more than a gratuitous, simple affirmation, a sort of concession to our spirit of " logic ", an answer which conforms with the " accepted scientific principle ".

So in the matter of the red-hot cast iron stove, the explanation given was that cast iron in a red-hot state is porous, and this was how the carbon monoxide came out of the stove instead of going up the chimney. If I had argued that carbon monoxide could not be present because of the fast combustion, it would have been said that the carbon dioxide, when passing through the cast iron, had taken on an extra carbon and become carbon monoxide. However, that would have resulted in decarbonisation of the iron, and I never saw the several hundred grams of carbon (about 40 grams per kilogram of cast metal) disappear from the stove and transform the cast iron into steel! Besides, those several hundred grams of carbon would soon have burned. Furthermore, even if cast iron is porous when red-hot, and even if carbon monoxide is produced by contact of carbon dioxide with the iron (another explanation which is sometimes given—and which I do not believe because if it were so produced it would immediately burn up to reform carbon dioxide) this would assume conditions which in fact are unrealistic. When the stove draws well, there is reduced pressure inside it, so that if the cast iron is porous, there would be no passage of the gas through to the outside; on the contrary, there would be an indraught through the porous wall!

Let us look at what really happens. It cannot be denied that in a badly ventilated room a red-hot stove can give rise to poisoning from carbon monoxide, which may even have fatal consequences.

The explanation only came to me indirectly, and by then I was over fifty years old, whereas when I was a child, the problem had impressed me by the contradictions in the teachers' explanations—

17

and, later, even those of professors at college—who had never " sensed " the contradictions.

(3) WITH THE WELDERS

In 1935 I made an observation which left me puzzled. A case of fatal poisoning of a welder had occurred, and my duties obliged me to take charge of the inquiry into the accident. This was carried out with the aim of determining the causes of the accident and preventing any recurrence. I could find no evidence to show where the carbon monoxide had originated.

On several subsequent occasions, similar accidents occurred, and at no time could I trace the source of the carbon monoxide which was supposed to have been inhaled. Because of my insufficient knowledge, I was forced to admit my ignorance of the exact causes of the accidents, but the facts remained subconsciously in my mind.

It was only in 1955 that it dawned upon me what had happened. In that year, in one district of Paris, three welders using blow-pipes had died in a period of several months. My collaborator E.P., responsible for safety conditions of workers in the particular district, had given me detailed reports. These, along with the autopsy findings from the public Prosecutor's office, made it quite plain that the welders (all three were oxygen-blowpipe cutters) had died from carboxyhaemoglobinaemia (carbon monoxide poisoning) and not from nitrous oxide poisoning.

In samples of the air taken from the work-places, there was no evidence of a dangerous carbon monoxide content.

It was decided, in collaboration with factory doctors of the firms where the victims had worked, to take blood samples from fellow workers even though the men were apparently in good health. The samples showed that those doing the same work as the victims were seriously afflicted with chronic carboxyhaemoglobinaemia, some to a degree approaching that of the fatal cases.

I decided to make a more extensive inquiry which, in fact, lasted for four years. Cases of carbon monoxide poisoning were continually occurring during this particular welding operation, but analyses of samples of air taken near the nose and mouth of the workers always showed that no carbon monoxide was being inhaled.

When my collaborator told me the results of his first investigations, he wrote saying that he hesitated to report such contradictions. He did not want to say that the workers were poisoned by carbon monoxide and to say, at the same time, that they had not breathed in any of the gas. If this were in the report, he could imagine the smile on the listener's face, a doctor perhaps, and one would be taken for a fool. He had no wish to send in reports which would only be received with a shrug of the shoulders and might well compromise his career.

I told him that on the contrary I had had similar findings over the last twenty years. I said it would be useful if he continued his inquiry and gave me figures of the carbon monoxide content in the workmen's blood and in the air around them. I also told him that I attached a great deal of importance to his reports; they would be carefully studied and not just filed away to end up as material for statistics.

In fact the three fatal accidents of 1955 had led me to a hypothesis which I had to verify. As the blood contains carbon monoxide without any having been inhaled, if there is an undetected source of this toxic gas, it would be found in samples taken in the proximity of the respiratory organs, thus carbon monoxide would be produced within the body—but from what?

The other important observation was that the most serious cases of poisoning occurred in men using oxygen flame cutters. The blowpipe itself was not responsible: combustion here produces carbon dioxide because there is an oxygen intake which burns up completely. Then there was the indisputable fact that carbon monoxide did not enter the breathing passages, but that the workers bent over the metal as they cut it, and the powerful flame jet made a large area of the surface incandescent.

Therefore, it was my opinion that the air, having been in contact with the incandescent metal, had become "activated". When the air was breathed in, it provoked a formation of carbon monoxide in the blood at the lungs level.

In order to confirm my supposition, an investigation was carried out in workshops where metals were heated to incandescence, which thus formed large surfaces in contact with the air breathed in. The result was the same even when heating was by electricity (resistance or induction); the workers were impregnated with carbon monoxide. This was confirmed by enlisting the help of several official laboratories in order to vary investigators and analytical methods. A continuous automatic recording instrument used in the investigation was allowed to run day and night.

A control experiment was made with the welders themselves. They were asked to wear a sand blaster's safety helmet, which has an air tube at the back of the neck. The tube was not attached to a compressor, but allowed to hang down so that the workers breathed in the air behind them. In a short time the incidence of carbon monoxide poisoning had markedly decreased. Therefore the source of the trouble had been the air striking the incandescent metal. Furthermore, a man standing beside the welders, but not leaning over the glowing metal, was unaffected.

As a result, the prevention of such accidents followed. All that was needed was to give the workmen an upward current of fresh air to counteract the air rising from the incandescent metal. A system of ventilation which sucked up the air would have on the contrary, a noxious effect.

Chemically there was nothing abnormal about the air in question which had always been a mixture of oxygen and nitrogen. Since carbon monoxide is composed of carbon and oxygen, I was led to consider the possibility of a transmutation of nitrogen into carbon. I was told this was "revolutionary" or "impossible", but I had already made other observations, to which I shall refer later, which had led me to believe that it might not be absurd to think in this manner, in spite of established science. I immediately realised that such an idea would furnish a clear explanation of certain mysteries which my subconscious mind still retained.

I then did some reading to find out if similar observations had been recorded, and I discovered some interesting documents.

I found reports from a number of countries of cases of carbon monoxide poisoning among blowpipe operators working on sheet metal (welding, oxy-cutting, sheet metal shaping which the blowpipe brings to red-heat). Samples of air had been taken which showed no noxious level of carbon monoxide.

This was astonishing because of the apparent contradiction. The English, and especially the Germans, had gone to great lengths to discover the cause of these accidents. The Germans had built a sealed chamber of 100 cubic metres capacity, in which powerful blowpipes heated metal red hot, but their measuring instruments recorded no trace of carbon monoxide after several hours. The experiments in England were made inside a closed hangar of an aircraft-carrier and confirmed the German result.

As this phenomenon occurred in the working of ferrous metal, and in view of the negative results obtained in the above experiments, the European Coal and Steel Community, with its vast financial resources, decided to undertake an investigation. This was also a failure and the medical director of the investigation, in which doctors and chemists cooperated, had to announce that the chemists were unable to detect traces of carbon monoxide in the air, but that it was present in the blood. The investigator was doubly at fault: first, because he admitted, in effect, that chemistry, alone, was tied to biology; and secondly, his verdict was unfair to the chemists. They could not find carbon monoxide which did not exist in the breathed-in air, but the doctors found it in the blood by chemical methods (physical methods could also have been used, and applied to the breathed-in air.) The techniques, therefore, were not to blame.

(4) AN EXPLANATION

Thus, a number of observations made in various countries, with the support of large financial resources, confirmed our findings that poisoning occurred while working with blow pipes without the latter producing carbon monoxide. But I had also established that production of carbon monoxide could be endogenous, and occur after the inhalation of air which had passed over an incandescent ferrous surface. (Experiments made in 1964 on rabbits and on humans had shown me that this endogenous reaction only takes place when the metal is heated to more than 400°C.)

I know that various explanations were advanced to account for this phenomenon after I had reported it to the Conseil d'Hygiène de Paris, but all the researches made by others could not disprove my explanation. A variation in oxygen pressure due to heat is not responsible. Methodical investigations of this possibility have been carried out, notably by Professor Desoille, which show that the phenomenon is independent of oxygen pressure. Nitrogen is the only explanation. Furthermore, if in the mixture of gases inhaled (nitrogen + oxygen), nitrogen is replaced by another gas such as helium which restricts the intense action of oxygen, the phenomenon does not take place.

Free nitrogen is never found in an atomic state. In air it exists in a molecular state—that is to say, there are always two nitrogen atoms bound together with their nuclei 1.12 Angström units apart. This unit is abbreviated as Å, which represents $1/10,000$ of a micrometre (1 micrometre or micron = $1/1000$ of a millimetre = 1μ). Therefore 1Å is also $1/100,000,000$ of a centimetre or 1×10^{-8} cm. As the Angström unit is not a homogeneous measurement, there is a tendency to replace it with a subdivision of the metre, called the nanometre or $1/1,000,000,000$ metre = 1×10^{-9}m., abbreviated to nm. Therefore 1 nm = 10^{-7}cm and 10 Å. Instead of nanometre we can also say millimicron, $m\mu$.

The electrons constituting the second layer of each atom—when the atom is isolated—are arranged in a molecular orbit round the two nuclei. (In the carbon monoxide molecule the two nuclei are a little closer together at 1.09 Å.)

I shall not deal with this phenomenon in detail (I have done so in other publications) and will only point out that if by the application of an energy (in the above case heat, although that alone is insufficient, there must also be the catalytic effect of an incandescent ferrous metal) the two nuclei of nitrogen combine, there is then one compound nucleus containing the nucleons of the two nitrogen nuclei. Each nitrogen nucleus has 14 nucleons, so the paired nuclei have 28. These 28 nucleons have a tendency to arrange themselves into two groups; one of 12 which is carbon, and the other of 16, which is oxygen:

$$12\,C + 16\,O \; :=: \; 2.\; 14\,N.$$

This preferential arrangement is perhaps due to the natural tendency of nucleons to divide into an association of two protons to two neutrons. Thus in the helium nucleus there is an alpha ray, or $\propto$ -particle, as in natural radioactivity. In such a case the pairs of nucleons are expelled (with a relatively low energy on the scale of nuclear physics). Therefore, a naturally occurring portion of these nuclei, and this portion alone, is spontaneously expelled. (This " primary " grouping in ∞ nuclei was accepted by Pauling in 1966.)

Now, carbon 12 has a nucleus composed of 3 alpha particles;

21

oxygen 16 has a nucleus composed of 4 alpha particles; the double nitrogen nucleus breaks up into whole alpha particles and not by division of one alpha nucleus (is this the reason why atomic nitrogen is so difficult to obtain?). See Fig. 6.

This is only a hypothesis, since we have no means of "seeing" what goes on inside the nuclei, but this transfer of nitrogen into an oxidised compound of carbon fully explains certain observations on the nitrogen cycle in plants, and the role of nitrogen in animal life.

It should be noted that the above phenomenon does not involve a "massing" or "fusion" of the two nuclei of nitrogen, because when the amount of energy is increased the nitrogen nuclei are so intimately joined that they constitute a group of 28 nucleons, of which 14 are protons, and this is silicon. So it can be seen that there is nothing to prevent us thinking of a bond between nitrogen and silicon, which opens up new perspectives on the origin of the earth's crust. In my other publications, I have cited a number of supporting results. (Figs. 10 and 15)

I have mentioned certain analogies between the molecule of carbon monoxide CO and that of nitrogen N_2. Both are isosteric molecules (having the same number of electrons). Here are some other similarities:

	CO	N_2
Melting point, °K	66	63
Boiling point	83	78
Density in liquid state	0.793	0.796

Viscosities very much alike.
(°K is degree Kelvin, or absolute zero, with a scale commencing at a zero representing a temperature of -273°C. Therefore 66°K corresponds to $-273 + 66 = -207$°C)

The spectroscopic method confirms the similarity in structure between these two molecules. The distance between their individual nuclei, though varying according to whether the CO or N_2 group are isolated from, or included in a more complex molecule, can be the same. It is always very close, being 1.12 Å for N_2, where the molecules are isolated, and 1.09 Å for CO. So the nuclei of CO are closer together, that is to say, more energy was needed before N_2 could become CO.

One can also suppose that this external energy is no more than an addition which tips the balance of a structure to the limit of stability. In the oscillatory movement of nuclei in the N_2 molecule, activated by external energy, there can be a metastable situation where the precarious internal equilibrium of the nuclei is broken, the nitrogen being at that point at its limit of stability. It is very difficult to obtain an isolated nitrogen nucleus (nitrogen is only stable in even numbers, hence the molecule N_2). The isolated nucleus has the same odd

22

number of protons and neutrons, and no heavier element is stable under such conditions. When a "peak of resonance" is reached between the two N_2 nuclei, one wonders if there is not also a "liberation" of internal atomic energy which could maintain stability. Thus a proton-neutron group (or deuteron) of one nucleus would move towards the adjacent nucleus, and so stability could be re-established. The donor nucleus would then become carbon (3 ∞) and the recipient nucleus oxygen (4 ∞). (See also chapter V, (3) for the inframolecular reaction, $N_2 \rightarrow CO$.)

Chapter 3

OBSERVATIONS ON PLANT ANOMALIES

> "What is new is always regarded with suspicion. But when one sees a gleam appear one has no right to put it out. One's duty is to study."
>
> P. COURRIER.
> Permanent Secretary,
> *l'Académie des Sciences.*

(1) FALLOW LAND

Agriculturalists have long known that virgin soil, never previously cultivated, can grow good crops without manure or fertilisers. However, its fertility gradually becomes exhausted as the soil becomes deficient in those elements removed by the crops.

Since time immemorial it has also been known that after a soil has been rested for a few years it becomes reconstituted: the missing elements have returned. This practice is termed fallowing. In systems of crop rotation a fallow period is arranged and this is often achieved by sowing leguminous plants, such as trefoil and lucerne, which "fix" nitrogen, thus shortening the fallow period and providing pasture at the same time.

Agronomists explained this by saying that the missing elements were brought in by dust, animals, by transfer in the soil, etc. A simple statement without having measured these apparent additions was that the elements reappeared by a progressive dissolution of the insoluble soil compounds. In order to expose this fallacy, we have given the total amounts of soluble and insoluble soil materials. The idea of solubility is too arbitrary: an insoluble compound can be rendered soluble by the secretion of root terminals, micro-organisms, etc, but it is a mistake to separate the soluble from the insoluble in the laboratory. When we speak of deficiencies, it must be understood that we mean a deficiency in any molecular form as well as in the soil contents.

(2) ACTION OF PLANTS

However, this process set up by agronomists had long ago been questioned.

About 1600, a Flemish chemist, Jean-Baptiste Helmont, had planted a tree in a pot containing 200 lb of soil. After a period of five years, he found that the tree had gained 164 lb and that the soil was also heavier, but by 2 oz only. He had not given the tree any mineral elements, only water, and thereby sought to prove that the water had turned into solid matter.

This summary experiment tells us little, but it does show that the origin of matter was a subject for research.

More precise investigations did not come until much later. At Nantes in the mid-19th century, Grandeau, after making certain soil analyses, showed that a soil left fallow tended to bring itself back into balance: if it were too acid or too alkaline, it returned to neutral of its own accord. This was confirmed by Garola at the end of the 19th century.

In Germany, Rudolf Steiner undertook research into this subject and, in 1925, founded " a school of balanced agriculture " in Switzerland. This activity was continued by Pfeiffer, a Swiss, whose writings have been published in French; the most important to concern us is " Fécondité de la Terre ". The following are some of his observations.

(3) LAWNS AND DAISIES

To have a good English-style lawn the soil must contain lime. When the lime is exhausted, daisies make their appearance and the gardener knows that to improve the lawn he must correct the soil. The greater the lime deficiency, the more abundant are the daisies.

Pfeiffer analysed the incinerated ash of daisies, and found it to be rich in calcium. He asked where it came from, since the daisies grew when there was no more lime in the soil. He could find no answer.

Obviously, one could not say it had come by migration in the soil, because then it would have been present in the lawn as well. The orthodox view was that daisies selectively " fixed " the lime; but they grow quite well when there is no longer any lime in the soil (providing there is silica along with suitable microorganisms).

For Pfeiffer, this showed the soil's natural tendency to re-establish equilibrium. When lime is lacking, silicon-loving plants grow, and their ash is rich in lime. When their stems and leaves die off in the autumn they bring to the soil the missing lime, and after several years equilibrium is re-established. There is a small growth of lawn grass (due to shortage of lime) and an abundant growth of daisies supplies the lawn with the lime it lacks. Such examples of plant associations are common in nature, but a mystery remains: how does the lime get into the daisies? We will see the answer.

(4) OTHER OBSERVATIONS

Pfeiffer's writings contained many similar questions. Here is a typical one:

" Buckwheat has a marked affinity for sand or silica, yet is characterised by its high lime content ". Wheat likes a soil relatively rich in lime, but incineration of its straw has yielded, for one soil, 6% of ash (relative to weight of dry straw) with a content of 5.8% lime and 67.5% of silica. On the other hand, when trefoil was sown with the wheat in the same soil, the trefoil, which prefers silica soils, had an ash content of 35.2% lime and 2.4% silica.

The content of silicon and calcium in most plants is unrelated to amounts of these elements in the soil. The composition of a plant is constant for each variety—at least this is broadly true, but quality can vary greatly with the soil, for example: celebrated wines which are quite localised; a cultivated digitalis without digitalin; a parsley totally devoid of vitamin D, etc.

The oak is a tree of granite and schist regions (soils rich in silica where lime may be totally absent), but the tree can have large amounts of calcium in its wood and bark (up to 60% lime in the ash).

Simoneton has repeated an already well-known experiment to show that geranium cuttings root well in pure silicon sand (Fontainebleau type) if supplied with rainwater, or even distilled water, without organic or mineral supplement apart from the silicon. A subsequent analysis of these plants shows that they have " produced " calcium and certain other elements. (Fresh sand with its bacteria is obviously essential, sterilised sand will not do.) However, it must be remembered that this so-called "pure" sand also contains, apart from silicon, traces of seven elements, representing 0.17% of its total weight. Of this 0.17%, oxides of iron, titanium, calcium and aluminium constitute 0.15%.

There are many such anomalies. One plant, the Tilandsia, commonly known as Spanish Moss, (a Bromacea), will grow on copper fibres without roots or contact with the soil. Its ash contains no copper, but has 17% of iron oxides in addition to various other elements which could not have come from the rainwater supplied to the plant.

One could cite such plant anomalies at length. Numerous studies have been devoted to them, and no explanation has been found, but by means of biological transmutations all these anomalies can be understood. This had already been foreseen by some during the 19th century; in 1880 von Herzeele acknowledged that " the findings on the creation of matter are not new ".

Chapter 4

ANOMALIES IN MINERALS

> "Our mind has a natural tendency to reject what does not come within the framework of the scientific expectations of our epoch. Scientists are men after all. They are impregnated with the prejudice of their class and times. They readily believe that what is not explainable in current theory does not exist."
>
> A. CARREL.

GENERALITIES

Unexplained phenomena have long been observed in stones and rocks, and eminent geologists such as Jean Lombard, the brothers Choubert, the professors René Furon, André Cailleux, etc., have contributed to these observations.

It is known, as a result, that dolomite rock (dolomite is pure magnesium carbonate) is formed inside limestone rocks (calcium carbonate) to the detriment of the latter. (Dolomite rock is a rock formation which contains a mixture of these two carbonates)[1]. To hide our ignorance of what took place one word has been used (as if it could be explained in one word); the word is "metasomatosis" of dolomite (or alteration of the soma or body). (One simple body, calcium, has changed into another simple body, magnesium.) Indeed we are back where we were in previous centuries, when the action of opium was explained by its "sleep-inducing ability".

It has also been observed that in very compact and absolutely impervious slate schists, "kidneys" of lime termed geodes, the size of a man's fist or larger, are sometimes present. The "explanation" given has been that the lime got there by migration. But from where, since there was no limestone for miles? And to have been carried by water the dissolved limestone would have to penetrate the schists, and these, as has been proved, are absolutely impervious to moisture. It should be noted that certain schists, probably under the effects of pressure which occurred later on, as in the Alps, have produced layers of slate inside which limestone has begun to form. This appears as an intimate mixture, and its formation certainly is endogenous. There could have been no diffusion of limestone-carrying water into these

---

[1] All mineralogists are not in agreement; some consider that $MgCO_3$ is magnesite and that the pure mineral dolomite is $Ca-Mg(CO_3)_2$, and that dolomite rock is an impure mixture of these two carbonates and of calcium carbonate, which would be calcite + magnesite + dolomite.

27

beds of impervious schists. This particular form of slate inside the schist is identified by the fact that it effervesces in acid.

(1) "ILLNESS" OF STONE

Sometimes the siliceous stone of monuments swells, and it can be seen that a layer of gypsum is formed, which finally disintegrates, and this is why sculptured figures on monuments disintegrate. (Quite frequently it is carbonate of lime that is formed.) The classic explanation is always the same: the sulphur in the gypsum (calcium sulphate) comes from the air, the soil, birds' excrement, etc., the calcium results from transfers, which concentrates on the surface as there is always some calcium in stones, so they said (without having analysed the stone). The trouble with this type of " reasoning " is that analyses have shown that the weight of calcium thus deposited on the stone's surface (either in the form of sulphate or carbonate) is notably greater than the total calcium content of the original " normal " rock.

One remembers that the operation of cleaning Notre Dame cathedral in Paris was postponed after it had been noticed that a siliceous crust had formed on the surface of the calcareous stone of the building, under the visible black layer, and had protected the underlying stone.

I should add that it is not at all certain that the black layer on the calcareous stone of the Paris buildings is due only to soot. Since a crust of silicon is formed on the surface through activity of aerobic microorganisms, one can envisage the reaction: calcium-carbon = silicon. In other words this formation of silicon would be accompanied by an " extraction " of carbon from its combination with calcium in the stone, hence the simultaneous appearance of the outer black layer of carbon and the underlying layer of silicon.

This uniform black layer can be observed even on buildings erected during the last century. There are many specialists who think that the impure atmosphere of large cities is not the only cause since the same phenomenon has been observed in rural areas. I have noticed, too, that the few Parisian granite monuments (for example in Place Fontenoy) and granite plinths do not have this black layer. There is therefore a microbial " disease " of limestone as well and not only a deposit of atmospheric origin but, in certain cases, granite can also be " ill "—see photographs of the church porch of Sizun, Finistère, (Brittany) in *Natural Transmutations*.

I should say at once that changes in rock are due to microorganisms: fungi, often of the family Aspergillacae and other moulds; microscopic algae; bacteria; actinomycetes such as streptomyces which live like bacteria but reproduce like fungi. In these phenomena there is therefore an interaction of living organisms followed by various chemical reactions between the elements set up by biological transmutation.

Does this mean that only living bodies such as microorganisms are capable of having changed the earth? No. In *Natural Transmutations* one sees how granite gives rise to schists by transmutation of elements, through the action of imperfectly known physical forces (probably temperature and pressure combined), and on an energy scale unrelated to biological forces. The latter, under conditions of low temperature and pressure, and with the large catalytic activity of the cell enzymes (those of animals, plants and microorganisms) are also capable of producing the same transmutations. By comparison, very high temperatures or pressures are needed to obtain a chemical combination of nitrogen and oxygen in the laboratory, whereas the living cell carries out this reaction at ordinary temperatures.

Hitherto, we have not been able to devise a simple and sure method for producing these transmutations. We have only been able to verify them *in vitro* by biological means, even in the case of minerals, metals and non-metals. This is why we will deal only with biological transmutations in this book.

(2) SILICON AND CARBON BONDS

The above will not prevent the reader from reconsidering the problems pertaining to the classical sciences, mineralogy, geology, pedology, agronomy, etc.; and at the same time realising that certain previously suggested explanations could be modified.

We have proposed the reaction:

Silicon = carbon + oxygen

The reaction illustrates the possibility of finding carbon in silicon (by the action of a still-unidentified energy) while rejecting the theory that carbon can come only from organic substances. Graphite is found in siliceous rocks so old that they probably antedate all vegetable life on earth. No one has been able to establish a connexion between the structure of graphite and that of any form of vegetation. On the other hand the structural arrangement of graphite in parallel layers is quite similar to that existing in certain clays, but while in a clay such as montmorillonite the particles are about 9.5 Å apart, in graphite they are no more than 3.35 Å, or about three times closer together.

It should not be concluded from this that graphite is a compressed clay, because clay also contains aluminium (which is silicon minus an atom of hydrogen, therefore boron + oxygen, and boron is carbon minus hydrogen).

But there is an analogous process, where the layering takes place only in silica rocks (which could be granite), and not on aluminium silicate. Coal deposits also exist in silica rock, forming bands or strata which sometimes contain up to 40% silicon compounds (an indication of—we believe—incomplete "metasomatosis", perhaps due to a lack of pressure?).

This is only one example disproving the assumption that all rocks, meteorites etc., containing carbon must have a constituent originating from an organic substance. The diamond, too, is certainly not connected with vegetable carbon and one must reconsider the formation of coal within schists, which practically always exists near a tectonic fold of the Primary and especially the Hercynian geological era. This by no means suggests a so-called tropical forest; the folding would have upset the earth's surface along with its vegetation, and this enclosed plant life would have left traces.

The presence of sulphur in some coals should be no surprise, and we shall discuss a probable origin of this sulphur, later.

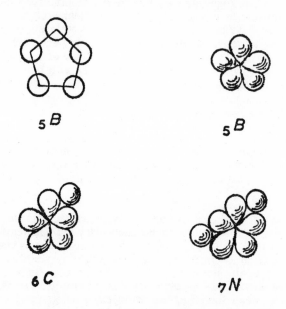

Fig. 7. Simplified scheme of atomic nuclei: only the protons are shown. A first layer would have (*above*) at the most 5 protons = boron. On a second layer (*below*), with one or more protons, is carbon $_6$C, (*left*): with two more protons, on the second layer, is nitrogen ($_7$N).

(3) CONTRADICTIONS ON THE ORIGIN OF LIMESTONE

Let us refer to the contradictory statements of those who have maintained that limestone appeared on earth, especially in the Secondary era, through the activity of molluscs. It is claimed that these creatures grew calcareous shells by "fixation" of the calcium from sea-water.

There are indeed shells to be found in limestone deposits, but in such deposits the shells have been "drowned"; one often finds traces or the shells themselves, of all sizes, in large numbers, sometimes scarcely visible to the naked eye. However, there is also an amorphous or a crystalline limestone "cement" surrounding them. And here it is difficult to attribute this "cement" to the shells which, because their traces prove it, antedate the substance that shaped them. This substance consists of a form of limestone which cannot derive from that of the shells (the "cement" is due to Coccolithes or Nannoconnus).

This should suffice to show that limestone is not solely produced by shells.

Another important consideration is that calcareous deposits may be as thick as 1000 metres (=about 3000 feet) or more (sometimes 2000 to 3000 m). If it is assumed that a concentration of limestone has been brought about by shell molluscs from calcium in the sea water, a calculation reveals the impossibility of obtaining such deposits, because sea water contains only 0.042% calcium.[1] Even supposing that at the end of the Primary era the calcium carbonate in the sea was at saturation point (and where would it come from in the sea as there is no trace of it in either the Primary or in the Pre-Cambrian eras—except some isolated and rare formations), the level of the sea can be estimated not to have changed by more than a few hundred metres since then. The average thickness of the ocean water layer is about 3600 m (=10,800 feet); and one can calculate from this the thickness of a calcium carbonate deposit representing the "fixation" of all the carbonate in the sea (assuming it to be the same proportion to the bottom). Given the very low solubility of this calcium salt in the sea water—rather poor in carbon dioxide—it is probable that the calcium carbonate content of the sea has changed very little since the beginning of the Secondary era. One can see how wide of the mark one is. Orthodox teaching again shows the naïveté of certain assertions which could well have been avoided by a simple calculation, even supposing the use of sedimentation flasks and a new supply of water. A precise calculation is impossible, because we do not know the movements of sea water, which depend on sea bed contours (there are many deep holes), salinity, temperatures and therefore varying densities. Underwater currents flow between largely stationary masses of water, or are subject to slow transgression. Consequently it would be wrong to assume that certain deeper areas had received the lime of all the incoming ocean waters. In the same way it is also a mistake and quite naïve to calculate the dilution of a noxious product put into the sea by dividing its volume by the volume of the ocean. (By way of reference, we can say that the total salts of the sea would make a layer 153 metres (=about 459 feet) thick if redistributed on the continents, and calcium would represent only about a ninetieth part of this layer. All the calcium in the sea could not, therefore, explain the amount which is on land.)

We believe that the limestone may have been formed by siliceous mud deposited in the deeper areas where they formed calcareous

[1] It would even appear that this rate is that of the superficial layers as deeper in there is none. (F. Rinne, *La Science des Roches*, Lamarre, Paris 1949)

sediments by the action of various microorganisms, or by the effects of physical forces whose nature can only be guessed. The lime then has covered the shells of those marine animals which had already formed their shells from the magnesium in sea water. A. de Cayeux in his book *30 millions de siècles de vie* (Thirty Million Centuries of Life)[1], states that from the Pre-Cambrian era to the present time the earth's magnesium has decreased in the proportion of 12 to 1. What has happened to it? We shall see that it is essentially magnesium which is utilised by crustaceans and marine animals to make carapaces and shells, corals etc. (Similarly microscopic algae such as "Lithothamnium" manufacture their own lime. These organisms are found around the French coasts in forms like coral, and grouped in reefs always immersed in agitated, oxygenated water.) In the Mediterranean the Corallinaceae are fixed on the shore at sea-level.

Certainly, the sediments in the ocean could not be limestone because the latter did not exist in a marked fashion on land. The marine sediments were necessarily compounds of silicon, of aluminium silicate, and in the deeper troughs of the ocean, by means of various forms of energy, they were able to produce mixtures of lime and clay, lias and marl. Later, as the transmutation of silicon into calcium proceeded, the calcium, carbon and oxygen needed to constitute a carbonate $(CaCO_3)$ gradually appeared. This would be formed from silicon $(Si = C + O)$ and at the same time part of the carbon would have merged with the silicon to yield calcium, Ca. All the elements required for a chemical change into carbonates were therefore to be found together at the bottom of the sea, and obviously there could not be any atmospheric carbon dioxide action.

(4) AN ORIGIN OF SULPHUR—FORMATION OF GYPSUM

In other cases it is not the carbonate of calcium which is present, but the sulphate, (gypsum, which, heated, gives plaster of Paris). The formation of sulphur is explained by the fact that one reaction " splits " silicon into carbon + oxygen, and a second reaction takes the carbon away to combine it with the residual silicon into calcium (carbon + silicon = calcium) with the result that oxygen is left free. But local conditions may be such that part of the free oxygen turns into sulphur, the atomic nucleus of which would be a coupling of two oxygen nuclei (oxygen + oxygen, or

$$\overset{16}{\underset{8}{O}} \ + \ \overset{16}{\underset{8}{O}} \ := : \ \overset{32}{\underset{16}{S}}).$$

It is at a later stage that the chemical reaction takes place, when all

[1] Reprinted under the title: *3 milliards d'années de vie* (Three billion years of life) in the Encyclopédie Planète collection.

the constituent elements of calcium sulphate ($CaSO_4$) are together. What are the conditions for bringing these elements together? Is the formation of gypsum due only to geophysical forces or to a bacterial action? I do not have sufficient data for a definite answer.

However, the "manufacture" of calcium sulphate from silica by the agency of micro-organisms no longer needs to be demonstrated. The organisms responsible for sulphur formation have been isolated and studied. They are the *Thiobacilli*, of which there are several species. They require sulphur in order to live, and a very simple experiment will show that they "manufacture" this sulphur. Take a test-tube and rub its inside wall with a wad of cotton wool soaked in sodium thiosulphate—there will be only a minute trace of sulphur on the test-tube wall; then introduce a suitable culture medium and inoculate it with thiobacilli, which will proliferate and an increase in the sulphur content of the medium follows.

Gypsum is produced in this way on limestone of monuments; those of siliceous sandstone are also attacked by thiobacilli in association with other micro-organisms: moulds, actinomycetes, silicobacteria and sometimes algae. *Thiobacilli* by themselves cannot transform silicon. They produce the sulphur which is oxidised by other species, but the presence of other "specialists" in the conversion of silica to lime is also necessary.[1]

All of which illustrates the weaknesses in so many text books, and many courses of instruction, even in higher education, on matters concerning the "mineral kingdom".

---

[1] It is probable that these are Streptomyces which can only act along with silicobacteria; the latter are autotrophic, that is they "consume" SiO_2, taking oxygen from it and opening up the crystalline structure of silica so that enzymatic penetration by Streptomyces is facilitated. There are more than 400 known species, almost all are very toxic to humans.

Chapter 5

PROOF OF THE EXISTENCE OF TRANSMUTATION OF ELEMENTS AND AN OUTLINE OF THE MECHANISM OF THE PHENOMENON

> Verify the results yourself before condemning.
> P. COURRIER,
> Permanent Secretary,
> *l'Académie des Sciences*

(1) TOTAL PROOF

A simple proof requiring no chemical analysis is the increase in weight of mineral salts in a germinating seed which is supplied with nothing but distilled water. Take two identical batches of seed, germinate one group on filter paper kept moist with distilled water. After three or four weeks, incinerate both batches. There will be a higher mineral ash content in the germinated seeds. (The difference between the two varies greatly with the species, but the increase is always at least 4 to 5%.)

A more advanced study than the above has shown that more of the heavier elements are produced by transmutation with the oxygen and hydrogen of the water; for example, magnesium has decreased, calcium has increased. Part of the magnesium has disappeared because it has become calcium. The nucleus of the magnesium atom (24 nucleons) has joined up with an oxygen nucleus (16 nucleons) to give a compound nucleus of $24 + 16 = 40$ nucleons, and 40 nucleons constitute the calcium nucleus. It should also be noted that there are 12 protons in magnesium (this number characterises the element: if there are not 12 protons, it is not magnesium). There are 8 protons in an oxygen nucleus, so we have $8 + 12 = 20$ protons in the compound nucleus, and 20 protons are characteristic of calcium. (40 nucleons are not an individual characteristic, argon also has 40.) There are calcium isotopes with more than 40 nucleons, but they always have 20 protons: only the number of neutrons changes.

(2) MECHANISM AND DESCRIPTION OF TRANSMUTATIONS

Equal numbers of protons and nucleons must be present, simultan-

34

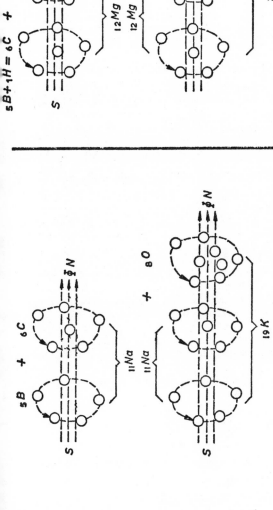

Fig. 8. Representation of nuclei (protons only) and transfer from one to the other.

First line: there is boron and carbon = sodium, to the left; on the right add 1 H/proton which goes into the boron centre to give C. We therefore get Na + H := 2 C + 2 C := Mg; therefore Na + H := : Mg.

Second line: on the left add O to Na; which gives Na + O := : K; on the right it is Mg + O := : Ca.

But one can see also that it is the representation of the left + 1 H in the left cluster; thus it is also K + H. There is no difference in the natural product, the representation of Ca is always the same whether Ca comes from K + H or from Mg + O. Likewise, as C + O := : Si, we see that Ca is also Si + C. Thus nature has three means of " manufacturing " the calcium which is needed. In the same way we can see that magnesium can come from sodium (with + H) or from calcium (with — O); that potassium can also be Na + O or Ca — H. (Φ=electromagnetic flux of N and S poles); the magnetic field due to the nucleus rotation is 165 gauss for iron; but the total inside field of the nucleus is 330 000 gauss in a magnetised iron atom due mostly to the electrons with very quick rotation—nearly 290 000 km/s for the K layer electrons, diminishing rapidly for the other layers to about 1 000 km/s for the external layer. For magnetised iron the spin axes of all electrons are parallel.

eously, to yield the type of transmutation we have demonstrated, at least in a biological reaction. (In the case of reactions producing a magnetic field, due to ionisation, it seems that a supplementary law may operate, but this will not be dealt with here, as we shall not consider reactions which only appear possible outside living organisms.)

This mechanism of transmutation is valid for isotopes, since in the reactions involved (which certain authors have termed the K-effect—Kervran effect) there is no expulsion of particles and there is no radioactivity found such as in the radioactive transmutations known in nuclear physics.

Let us take as examples the isotopes 16, 17 and 18 of oxygen (the only stable ones) and the isotope 23 of sodium (the only stable one).

$$\text{Sodium–23} + \text{Oxygen–16} = \text{Potassium–39}$$
$$\text{Sodium–23} + \text{Oxygen–17} = \text{Potassium–40}$$
$$\text{Sodium–23} + \text{Oxygen–18} = \text{Potassium–41}$$

There are no other natural potassium isotopes; isotope 40 is slightly radioactive, losing half its mass in 1.3 billion years. For mankind it can be considered stable. In sea water potassium—taken in chemistry as reference, there is very little natural ^{40}K isotope, only 0.01% whilst there is 6.89% of isotope 41 and 93.10% of isotope 39. The proportion of potassium isotopes can vary in biological reactions; in certain varieties of plants there is an increase of more than 15% of ^{41}K.

Referring to the above reactions, there are 11 protons in the sodium and 8 in the oxygen. These are irrespective of the number of nucleons in the isotope, or of its neutrons. The number of the latter is always the difference between the number of nucleons and the 8 protons.

Therefore there are always $11 + 8 = 19$ protons; this is the characteristic or " identity mark " of potassium.

This outline of the mechanism should make the reactions we shall now consider more easily understood. To comprehend certain other limitations, my other books should be consulted. There are many incompatibilities, and one must not simply "manufacture" non-existent elements in nature from written equations.

The addition of protons and nucleons of the two elements to a number equal to those of the new element provides the two main conditions required, though in certain cases these two conditions are not themselves sufficient.

A representation by chemical symbols explains the examples given above:

$$^{23}\text{Na} + {}^{16}\text{O} :=: {}^{39}\text{K}$$
$$^{23}\text{Na} + {}^{17}\text{O} :=: {}^{40}\text{K}$$
$$^{23}\text{Na} + {}^{18}\text{O} :=: {}^{41}\text{K}$$

36

We have used the sign $: = :$ easily made by a typewriter, as distinct from the mathematical sign $=$, which is used only in mathematics (or in chemistry to indicate two valencies) and the sign $\rightleftharpoons$ used in a chemical reaction to signify that it is reversible. If we wished to give the exact structure of a chemical reaction, for instance between Na and O, we write: $2Na + O \rightarrow Na_2O$. Two atoms of sodium, a monovalent, are needed to combine with one atom of oxygen, a divalent; the result is sodium monoxide. But it is possible to have other combinations since there may be two atoms of oxygen, each one " coupled " with a sodium atom. The molecule is sodium peroxide, Na_2O_2. There is a vacant place, a void in the peripheral electron layer of these oxygen nuclei called a free valency; similarly we can have CO and CO_2.

In a reaction of the type described, which directly involves the nucleus (a nuclear-biological reaction), we put $Na + O : = K$. This emphasises that it is not a chemical reaction and avoids confusion. The equal sign between protons and nucleons on the left and on the right thus indicates a reaction going from left to right. (Note that in chemistry the indices are placed below on the right: H_2O, H_2SO_4, NH_3 etc.) Reversibility does not take place as in chemistry, where an equilibrium is established between the atoms on the left (the components) and the compound molecule on the right (or in the reverse direction), whilst, in a nuclear-biological reaction, we write $: = :$, to indicate that the reaction is reversible, but not by the same agents. Different enzymes are required; certain reactions are only possible in one direction with animals, and in an opposite direction with plants.

Reactions can be subtractions, as with saltpetre or potassium nitrate formed on the surface of limestone walls. The potassium in this compound originates from calcium as a result of bacteria taking away hydrogen from the calcium atom: calcium $-$ hydrogen $=$ potassium, or $^{40}Ca - {}^{1}H : = : {}^{39}K$. The reverse reaction is $^{39}K + {}^{1}H : = : {}^{40}Ca$. This is why the sign $: = :$ is used.

So we can say that potassium has two origins: sodium plus oxygen, and calcium minus hydrogen. Thus, one of the most important laws of these biological transmutations becomes evident. This particular law governs many reactions with oxygen and hydrogen up to the nuclei level of atoms. In biology there are not only chemical reactions with oxygen (oxidation) or hydrogen (reduction), there is also a phenomenon of a much wider and deeper nature than these chemical molecular reactions which are merely " couplings " of peripheral electrons. (We shall see later that there are also reactions involving addition or subtraction of carbon; in the case of potassium, I have so far been unable to establish whether it can combine with aluminium; this would give

$$^{27}_{13}Al + {}^{12}_{6}C : = : {}^{39}_{19}K$$

but this reaction does not occur with all stable isotopes of potassium. For the isotopes 39 and 40 this reaction cannot be rejected *a priori*, but positive experimental research is needed before we can assert that a bond exists.)

Here a further statement is useful: often, to simplify the writing of nuclear-biological reactions, only the number of nucleons is given above and on the left of the symbol. This is because it is assumed that the reader knows the appropriate number of protons (which is the position the element occupies in Mendeléev's table, also called the " periodic classification ") since the number of protons is sufficient to identify the element. Therefore, if the symbol is Na one knows it is sodium, and there would be no point in adding the number of protons to identify it as sodium.

Naturally, we can write it if we wish. If we have $_{11}Na + _8O :=:$ $_{19}K$, we add together the protons written on the lower left of the symbols. If, to begin with, we have $_{11}Na$ and $_8O$, we know we have to find the element which has $11 + 8 = 19$ protons. The atomic number of potassium is 19; therefore sodium + oxygen give potassium.

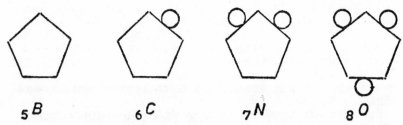

$$_5B \qquad _6C \qquad _7N \qquad _8O$$

Fig. 9. Simplified diagram: the core is shown by a pentagon representing five protons. Around the core are the protons of the second layer—$_5B$, $_6C$, $_7N$, $_8O$.

The same method applies to the other elements. The complete formula is:

$$^{23}_{11}Na + ^{16}_{8}O :=: ^{39}_{19}K$$

addition of the upper numbers on the left of the formula gives the upper number on the right, and likewise with the lower numbers. It is important to note that in these formulae the number placed top left is not the atomic mass, but the number of nucleons—therefore there are never numbers containing decimals, as in chemistry. In nuclear physics, the " mass " number is also used, which is the whole number nearest to the atomic mass. For example, for chlorine there are the isotopes 35 and 37, with approximately three quarters of the former and one-quarter of the latter existing in natural chlorine. In chemistry, elements are often mixtures of isotopes.

(3) A "MOLECULAR" EXPLANATION

The concept of "clusters" of nucleons which unite or separate while still remaining intact (that is, the group of nucleons forming a cluster does not change, and each cluster is the nucleus of an atom of a limited number of simple bodies) came to my mind for good reasons.

We have seen that what apparently takes place can be shown diagrammatically. Moreover, the concept shows that one does not "split" the basic nuclei, which are the "building blocks" for constructing the atom. The binding energy between protons and neutrons in the "sub-nuclei" does not change; the only energy involved is that which either separates the two nuclei or draws them closer.

The observations we have made show that only energies measured in keV are involved, whereas in nuclear physics energies are expressed in MeV, therefore a thousand times greater. (I would point out that 1 MeV = 1.6×10^{-6} ergs; therefore 1 erg = 625×10^3 MeV or 625,000 MeV or 625 billion electron-volts. These are therefore very low values, but they are applicable to a single atom; the energy of 1 MeV = 3.82×10^{-14} small calories.)

In chemistry the bonding energies are usually electron-volts, sometimes even less than 1 eV, which is 1000 times less than in the new science where we are establishing our first facts.

An appendix in *Low Energy Transmutations* gives part of a study made by the physicist L. Romani, showing that a drawing nearer of nuclei of a molecule could explain the mechanism we have proposed, and that the energy needed to set it in motion is at the rate of keV. This confirms our findings.

It can be said, therefore, that the reactions we have described constitute an infra-molecular phenomenon, a kind of "molecular fusion" in which two nuclei of a bi-atomic molecule are drawn together. Sodium hydride, NaH, through drawing closer of nuclei, would change to magnesium Mg; potassium hydride KH would lead to calcium Ca. We could also have $N_2 := Si$; $O_2 := S$; $C_2 := Mg$. The molecule MgO would become Ca etc. In addition there could be $N_2 \rightleftharpoons CO$, and we have already seen the common or very similar characteristics of these two isosteric molecules.

As a result of enzyme activity (as yet unexplained) one proton of one of the N atoms of the molecule N_2 goes to the nucleus of the other atom (with its associated neutron). In the place of the first N, there is now C, and the second N becomes O. Energy is required to complete these changes and by virtue of this absorption of energy, the nuclei C and O are more closely brought together than were the two N nuclei. There is a new stability at a higher energy level (Fig. 10).

But can this phenomenon, which produces endogenous carbon monoxide poisoning through the "activation effects" already described, be reproduced *in vitro* by physical energies measured in keV? We can conceive that under the effect of very high frequencies these energies come in resonance with the vibratory motion of the nuclei of the bi-atomic molecule, (which move close and separate like two balls joined by a spring. The wavelength of the C–O molecule is approximately two peaks of about 9 and 8 μ, i.e. a frequency of 3.3 to 3.7 $\times$ 10^{13}, or an average of 3.5 $\times$ 10^{13}. This frequency is characteristic of the " coupling " of the two nuclei and, since it comes in the infra-red region, it can be measured by an infra-red spectrometer[1]. Where there is a resonance, the " peak " of energy would be sufficient to transfer a proton (with its neutron) from one nitrogen nucleus to the other. As both revolve in the same direction we have, therefore, the phenomenon of electrodynamic attraction occurring just at the moment when the new nuclei are formed, without any changes in orbits of the electrons; the molecular orbits remain constant. Thus, the transfer of nitrogen to carbon monoxide would only be the transfer of one molecule to its isostere, with the nuclei remaining essentially equidistant.

Much greater energy is needed for a more important nuclear coupling, for example that of Na and H to give Mg, or Mg and O to give Ca. In such a case there is an inter-atomic action within the molecule accompanied by a break in one of the atomic K-orbits. The liberated electrons then arrange themselves in other orbits. There is only a single K-orbit for the two electrons (of opposite spin) which are placed very close to the two original nuclei which now are very near to each other. The original molecule disappears, to be replaced by a new element, but the total number of electrons remains unchanged.

NOTE: It is worth mentioning that in the case of $N_2 \rightarrow$ CO the molecular vibration is in the infra-red at 6 to 8 μ, and it is iron. quite close in the infra-red, which makes N_2 metastable—the wavelengths are very near to each other (the intensity is still considerable from 1100 to 1300° for 6 to 8 μ) from which a very slow impulse is possible, producing by resonance a metastable position. But now by what energy does this molecule become CO only in the organism? At the moment, the question remains unanswered.

It should again be noted that the total ionisation potentials (or energy necessary for displacement of all the electrons) are very similar for N+N and for C+O. Expressed in electron-volts this energy is:

$$N = 1473 \qquad C = 1025$$
$$\underline{ O = 2033}$$
$$2N = 2946 \quad C + \overline{O = 3058}$$

or an average of 3002±56.

[1] If C and O constitute a carbonyl grouping C=O in a complex molecule, the wavelength decreases by about 6 μ or a frequency of 5 $\times$ 10^{13}Hz, values characteristic of a complex construction. I give the wavelengths and frequencies universally recognised to avoid the irrational jargon of the specialists, who say that the vibration of C = O is 1060-1270 cm^{-1}. Calculations are necessary to express these in standard units.

Here again it can be seen that energy (112 electron-volts) had been necessary before 2N could become C + O.

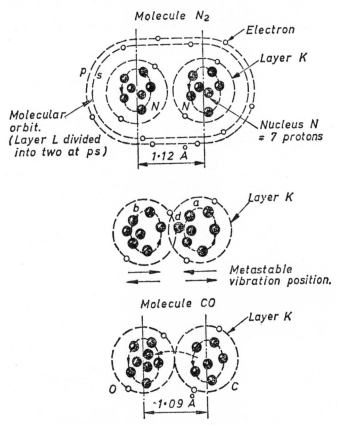

Fig. 10. Infra-molecular reaction, transfer of one proton inside isostere molecules. 1. The two atoms N of the N_2 molecule, with the protons representing the nuclei. 2. There would be resonance with the N_2 molecule's own vibration acting on the two internal protons (the others, linked by the electro-dynamic force, vibrate together), this under the influence of an electro-magnetic field of appropriate frequency, in the infra-red; the vibration amplitude of the internal protons, which are more free, means that, with resonance, the protons can leave the proton orbits a – b in such a way that, for example, the d proton is nearer the b orbit than the a; it is then in the electro-dynamic attraction field of b and is caught. (Only the nuclei and the layer K are shown.) 3. According to a law of exclusion stating that there cannot be more than five protons in orbit, d is electro-statically pushed to the centre: we find, therefore, O to the left, C to the right and the molecule which was N_2 becomes CO, without the electronic layer being changed (in the new equilibrium the nuclei are a little closer together).

(4) SOME DETAILS OF AN EXPERIMENT IN CHANGING SODIUM TO POTASSIUM

Let us now look at an experiment which has confirmed that potassium can be obtained from sodium in the presence of oxygen (which exists in air).

The experiment was made by one of my 'disciples' H. Komaki, a Japanese professor of science and director of a laboratory of applied microbiology. My book, *Biological Transmutations* had been translated into Japanese by N. Sakurazawa and came to Professor Komaki's notice.

In January 1963 he wrote to me saying that the transformation of sodium (yang) into potassium (yin)—according to the definitions of the Far East philosophical tradition—greatly interested him, and was of major importance to the economy of Japan, a country without potash deposits but with ample supplies of sea salt. (Using potassium they obtain four times the dry matter, with two rice crops annually.)

He told me that he and his collaborators would try to confirm this reaction and that he hoped to interest his students in the subject, with the idea of applying it on an industrial scale.

In November 1964 he sent me his results which I summarise below. He also told me that a large scale production was contemplated and that finally brewers yeast had been decided on as the agent for industrial production. As the dry matter in yeast contains 50% of edible proteins, two birds were killed with one stone: an industry was created and a valuable source of protein was developed.

A — EXPERIMENTAL METHODS

(a) Tests had led to the use of two species of mould from different genera, *Aspergillus niger* and *Penicillium chrysogenum*, and of two species of yeast, *Saccharomyces cerevisiae* (spore-forming) and *Torulopsis utilis* (non spore-forming).

(b) The potassium content was determined by the usual spectroscopic method.

(c) Culture flasks containing 200 ml of nutrient solution were sterilised for 10 minutes at a pressure equivalent to two atmospheres, to destroy any germs in the solutions.

(d) The quantity of potassium in the sterilised flasks was determined in order to ascertain the amount of potassium existing as impurities in the solution, and also that emanating from the glass (the latter was not inconsiderable since in the autoclave an exchange of ions had taken place with the ammonium salts). A similar check was made after the flasks had been agitated for 72 hours at 30°C, and revealed no detectable increase in potassium. This control made without addition of micro-organisms exactly reproduced the physico-chemical conditions used in the experiment. It was found that in a 200 ml flask the total potassium, both as an impurity and from the glass, was a little less than 0.400 mg.

(e) Czapeck's nutrient medium was used for the moulds, and Mayer's for the yeasts. (The materials were pure products complying with the requirements of the Japanese Industrial Standards. The water was purified by treatment with ion-exchange resins; potassium as an impurity was determined as above.)

(f) After an experiment with these accepted culture media, it was decided to assess the role of potassium in the cultures by utilisation of flasks in which the potassium salts of Czapeck's medium (for moulds) had been replaced by the corresponding sodium salts. Thus instead of one sodium salt and two potassium salts in the standard Czapeck solution, there would be three sodium salts. The solution therefore contained no potassium. Mayer's solution for yeasts was similarly treated, with the potassium salt replaced by its sodium homologue[1].

B—RESULTS

The sterilised flasks, with and without potassium, were inoculated with cultures of micro-organisms certified as pure strains by the Institute of Fermentation of Osaka. The 1 mg inoculum contained slightly less than 0.01 mg K.

(a) Potassium Increases the Yield

After 72 hours at 30°C the values for dry matter, centrifuged, (in mg per flask, flask shaken) were as follows:

With K	Without K	Species of Micro-organism
557.4	161.1	Aspergillus niger
906.6	189.6	Penicillium chrysogenum or P. notatum
1361.6	299.3	Saccharomyces cerevisiae (brewers yeast)
2634.9	460.0	Torulopsis utilis (wild yeast)

These differences are quite significant. The vegetative organisms in which potassium is available produce four to five times more weight of dry matter than those without available potassium.

(b) Comparison of Potassium Contents

The amounts of potassium in the dry matter (in milligrams per flask) were:

Cultured with K	Cultured without K	Species
5.35	0.901	A. niger
10.27	1.049	P. chrysogenum
15.84	1.749	S. cerevisiae
22.37	2.017	T. utilis

[1] Author's note: Here is the composition of Mayer's solution: Glucose, 10%; ammonium tartrate, 1%; $MgSO_4.7H_2O$, 0.25%; $CaHPO_4.2H_2O$, 0.08%; Na_3PO_4, 0.5%; distilled water to 100% or water purified to maximum on ion exchange resins.

It should be noted that these values represent the total amounts of potassium in the flasks. They are an average from a series of three flasks with K, and a series of 5 and 10 flasks without K. This information is interesting from an economical point of view. But experiments with non-inoculated control flasks showed only 0.400 mg per flask, which represented only impurities and K from the glass of flasks through ion exchanges with ammonium ions in the nutrient solution.

(c) *Without sodium or potassium—no vegetative growth*

A contrasting control experiment was made. The nutrient solution for yeasts and moulds was deprived of all potassium and sodium, resulting in no culture growth whatsoever.

CONCLUSION

The cultures in question are impossible in the absence of potassium and sodium. If only sodium is supplied, the organisms yield 4 to 5 times less dry matter, and the dry matter contains about half the potassium yielded in a nutrient solution containing potassium. The potassium produced is therefore about 1/10th of that produced in a solution where potassium is available. However, in every instance where organisms have sodium available, an inoculum carrying 0.01 mg K will produce 1 to 2 mg K, according to species, within 3 days.

NOTE: This conclusion should not be surprising. It reveals a phenomenon that I first reported in 1962 in my book *Biological Transmutations*, when I wrote of the need to "start" the reaction. As an example I cited the thiobacilli which need a small amount of sulphur for proliferation and for a subsequent production of sulphur greater than that supplied.

My explanation was fully supported by Monod and Jacob, the Nobel Prizewinners in 1965. Briefly, one can say that each enzyme is synthesised by a specific gene. There is a particular gene for the enzyme which induces formation of potassium from sodium, but this gene is inactive if blocked by an inhibitory agent whose action may be cancelled by a sufficient rate of K. Below this rate the inhibitor is active and blocks the gene, with the result that the enzyme is not synthesised (in man it is aldosterone that is synthesised).

It is evident, therefore, that experiments in biological transmutations may not succeed with highly purified substances. In the present case a certain initial amount of potassium is required; if it is absent the reaction is either incomplete or does not take place at all. This explains the difficulties reported by researchers who wished to use very pure substances, devoid of potassium, when attempting to demonstrate that potassium could appear although none had been present in the substances utilised.

Experiments must take this phenomenon into account. We should

44

not seek the presence of potassium where there is none but we should seek an increase of it in a solution on comparing it to a solution to which no micro-organisms have been added.

The decrease of manganese and increase of iron in leguminous seeds germinating in water supplied with soluble manganese salt has shown that results can be spectacular.

Komaki's experiments have been repeated under similar conditions in France. Using brewers yeasts in the nutrient solution, a known quantity of potassium was added and within three days a considerable increase in potassium was observed (the composition of the experimental solution was the same as that given with the addition of known amounts of potassium salt). To avoid any error, the total potassium in each flask was precipitated and weighed, instead of measuring the potassium by diluting the floating dry matter and by flame spectroscopy on an aliquot portion of the remaining liquid.

The above experiment was carried out by P.B.; details of methods and materials along with the statistical results will be published later.

I should emphasise that the foregoing experiments are not isolated cases; much earlier ones had been carried out on potassium production by fungi, but under different conditions. Here are briefly some figures from analyses made by V.H.

(a) Five fungi of the genus *Lactarius* are grown on the dry bark of a plum tree. In 100 g of this bark are 7 g $CaCO_3$, 0.1 g. MgO, 0.08 g P_2O_5, 0.133 g iron oxides, 0.113 g SiO_2 and traces of K_2O (dosing was done by gravimetric analysis to the nearest 1 mg).

In 25 g of fungal dry matter there are 0.025 g $CaCO_3$, 0.041 g MgO and 1.187 g K_2O. So there is approximately 49 times more potassium than calcium although there is virtually no potassium in the bark where the mycelium developed.

(b) Another experiment, this with fungi of the genus *Lepiota*. Cultivation on sand containing a small amount of clay; in 100 g of sand there is 0.006 g K_2O. The harvested fungi yield 36 g of dry matter, containing 1.570 g K_2O, which represents almost 300 times more K_2O than in the original sand.

(c) Another species of *Lepiota*, on the same sand, has yielded 16 g of dry matter and 1.210 g or 7.5% K_2O, or 200 times more K_2O than the 100 g sand.

Therefore these fungi which only proliferate for a few days, produce 200 to 300 times more potash than is contained in the growth medium.

H. Komaki has demonstrated that starting with 0.01 mg of potassium, in the case of yeasts it is possible to obtain a K average of 1.5 mg (or 1.8 mg of K_2O), or an 150-fold increase in three days. Rates of potassium increase by fungi of either simple or advanced biological structure are comparable. In the experiments outlined, controls ensured that no potassium arrived by migration.

It is also possible that certain fungi produce their potassium with

45

calcium as a starting point (only one experiment has been conducted to this end and awaits confirmation).

Take a wooden box 25 x 20 cm coated inside with paraffin wax. Pour 300 cubic cm of 8% glucose solution containing no potash into the box and add 1 gm calcium nitrate. Place the box under a bell-jar standing on a base of cotton wool to filter the air. Maintain the temperature at 20°C. After several days a surface layer of moulds appears. Experiments over a period of three weeks have given, after reducing the contents to ash, the following figures for three boxes respectively:

   0.059 g   0.064 g   0.095 g potassium sulphate

Control boxes, maintained under the same conditions without added calcium nitrate, had no mould growth.

So, initially there is no potassium, but after three weeks there is 60 to 95 mg K_2SO_4 or 30 to 40 mg of potassium in each box. (For dosing, the potassium was precipitated as sulphate.)

These experiments, though not as precise as those of Professor Komaki, have given evidence of potassium production by moulds which contain a higher quantity of potassium than they could have taken up from the culture medium. Their role is only that of clarification. It should be understood that Komaki's experiments are in no way final: an experiment is never complete in itself, there is always something more to be investigated. The result of Komaki's study has been to show that certain lower forms of fungi (some distinct species of moulds and yeasts) will produce potassium when they are supplied with sodium and do not multiply without sodium and potassium. His experiments had no other purpose: this is why neither the residual sodium nor calcium were determined at the end of the procedure (but potassium was found remaining in the solution, apart from the potassium in the dry matter). The aim was to demonstrate the fact, and not to measure the amount produced. On the other hand, further experiments on germinating seeds have included in certain cases quantitative determinations of 8 elements in the seeds and in control samples. In this way a quantitative comparison could be made of the increases and decreases taking place.

(5) TRANSFORMATION OF SILICA INTO LIME

We have referred to siliceous rocks becoming calcareous through the agency of micro-organisms. This is only one example. We have shown in our writings that this property has been known and utilised for a long time. Even in ancient times, horsetail (*Equisetum*) which is rich in silica, was used for recalcification. Silica yielded lime (formerly it was given to tuberculosis patients to hasten calcification of the lung cavities).

X-ray photographs have shown that bone fractures are healed much more quickly after administration of organic silica extracts obtained from horsetail than after giving calcium. Mineral calcium is a residue, and is not assimilated by the organism. In man and higher animals it exists in a terminal form, but plants and micro-organisms carry out the reverse reaction and utilise calcium. In recal-cification, therefore, mineral calcium should not be administered; instead, conditions must be established for the organism to "manu-facture" its own calcium. Nature has several ways of achieving this, and we will consider them later, but an important one is that involv-ing silica. In man it is organic silica (present in plants only at the period of spring growth) which must be utilised. Mineral or inorganic silica has a contrary or decalcifying effect. (Many uses for organic silica in recalcification have been employed under the direction of Dr. Charnot, in Rabat. In France, the subject has been dealt with especially in articles by Dr. Monceaux.)

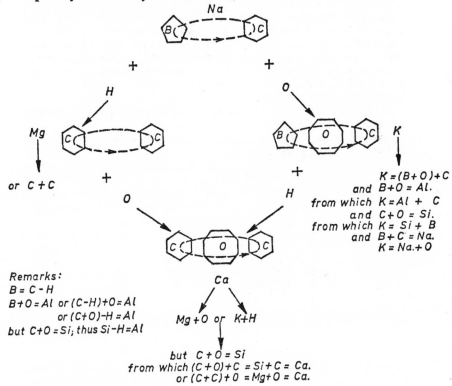

Fig. 11 shows the main biological transmutations. Each polygon represents a nucleus of a number of corresponding protons: 5 for Boron B; 6 for Carbon C; 8 for Oxygen 0.

NOTE: We see that Nature can rebuild numerous elements from a small number of "brick models"; the above reactions are examples, with known symbols, but do not include all the intermediaries; they are only explanatory formulas.

On the other hand, bacteria are able to transform the silica molecule (chemical formula SiO_2) and these silico-bacteria break down the molecule, taking up its oxygen. Such bacteria are autotrophic, that is, they feed on mineral salts in contrast to heterotrophic bacteria which live only on organic matter.

There are other micro-organisms which take up carbon atoms. They include moulds of the genus *Penicillium*—of which one species produces Penicillin—of the *Aspergillacae* family, but mainly actinomycetes. The carbon atoms are then bonded to the silicon (Si) atoms to give calcium.

$$\underset{14}{\overset{28}{(\ \ Si}} + \underset{6}{\overset{12}{C}} \ :=: \ \underset{20}{\overset{40}{Ca)}}$$

The first edition of *Biological Transmutations*, in 1962, referred to research being carried out to study the phenomenon more fully. In the 1965 edition we reported some results of this research, which formed the subject of a communication made to the Academy of Sciences in Paris on June 29th, 1964. The complete text can be found in Book 258, Volume 26, Section 13, pages 6573 to 6575.

Here are the essential points:

Our study concerned the sandstone of monuments in Cambodia, especially of those at Angkor (9th to 13th centuries).

The stone is composed essentially of quartz and of a somewhat rare clayey ferruginous feldspar.

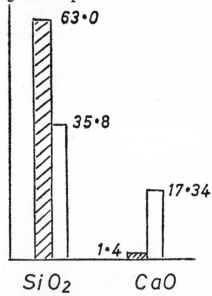

Fig. 12. Silica and calcium variations in healthy stones of the Angkor Vat temple (hatched column) and in deteriorated stones (blank column).

48

Analysis showed the following composition:

	SiO$_2$	CaO
Normal stone	63.0	1.40
Deteriorated stone	35.8	17.34
(Angkor Vat, external facing)		

In order to show that this more than 12-fold increase in calcium was not due to rainwater, the authors analysed the rainwater and also the water trickling at the lower west face of the monument. They made calcium ion determinations and found 0.8% in rainwater and 0.8% in that of drips. The latter in "washing" the facade had not therefore taken away any lime; it carried the same amount after as before.

It can be seen, therefore, as lime appears, so the silica disappears to a marked extent.

There is no need for a detailed analysis of all the elements; we had established that magnesium (with oxygen) could also give calcium, in the same way that potassium gives with hydrogen. These elements were then determined but with the following results:

	MgO	K$_2$O
Normal stone	2.50	1.20
Deteriorated stone	2.64	1.20

So the lime could not have come from these two minerals; no element in the stone had significantly decreased except silica, SiO$_2$.

An investigation was made of the micro-organisms present on the monument. Out of 120 samples taken, 83 contained actinomycetes, all of which were in the deteriorated stone, whereas there were none in the samples from the healthy stone. The authors reported identification of seven species of actinomycetes, all belonging to the genus *Streptomyces* (from one of which streptomycin is produced). "We have observed these actinomycetes to be much more abundant in the stones which showed most deterioration ... The almost constant presence of these micro-organisms in lesions of the stones permits the view that they have an important role in the process of deterioration."

In a paper, published by the Academy of Sciences, it is established that there is a marked increase in calcium (varying at 1158%) without any detectable amount of the element being supplied externally by water. On the other hand, the silica alone decreases in considerable proportion. This change in the siliceous stone and the accompanying increase of lime is linked to the presence of micro-organisms. If instead of giving relative values for SiO$_2$ and CaO we only consider absolute weight variations of Si and Ca, we find they are almost in equilibrium at 9.8 $\pm$ 1.5 (the variation is because there is no Si = Ca, but Si + C = Ca).

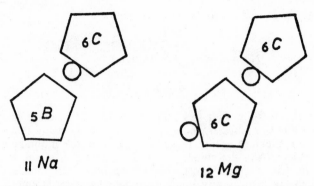

Fig. 13. The change-over of $_{11}$Na to $_{12}$Mg with one more proton.

These are only a few examples illustrating the research; there are others in the text which follows.

Chapter 6

PRINCIPAL TRANSMUTATIONS STUDIED

> In order to find out, we must be able to doubt.
>
> PASTEUR.

(1) THE SODIUM-POTASSIUM BOND

It has been shown that potassium can be produced from sodium. The reaction is very important in animal biology, and I have described it at length in previous publications.

In 1959 I was able to confirm the reaction in the course of an official assignment in the Sahara where I saw the working conditions of workers engaged in boring oil wells. I was sent on this mission by the then Minister of the Sahara, Jacques Soustelle, an ethnologist.

In the Sahara it was once thought dangerous to be exposed to the sun, and it seemed inexplicable that the oil men could now undertake strenuous work in mid-summer on unshaded metal platforms. A thorough investigation of the question was therefore undertaken by an army doctor and a team of assistants. Tests were made over a six-month period on a group of volunteers; everything they ingested and excreted was analysed (analysis of excreta was done in Strasbourg, the samples having been sent by air in refrigerated containers).

The results showed that in intense heat, excretion of potassium in the sweat greatly increases, but there is also an increase in ingestion of sea salt; the workers ate salted food and had salt tablets to suck.

However, not all the salt was excreted. What had happened to it? It was not stored in the body, since the difference between intake and excretion was such that an accumulation at this rate was impossible.

The greatest mystery lay in the heat or thermal balance. The number of kilocalories (kcal) taken in daily by one man from the sun's heat and from food reached a total of 4085 (average over 6 months). In summer the total was 7000 kcal. The sweat was 4.12 litres daily, which was evaporated immediately in the hot dry atmosphere. As 540 kcal are needed to evaporate 1 litre of water, there was a disequilibrium to such an extent that the workers should have died of hyperthermia (or heat stroke). Heat loss could only be achieved by sweating; the daily loss was $540 \times 4.12 = 2225$ kcal and the body received 4085 kcal. Therefore by the usual methods of balancing intake and excretion the daily excess was $4085 - 2225 = 1860$ kcal.

I came to the conclusion that the sodium, disappearing to become potassium, induced an endothermic reaction (absorption of heat); which explains why a person instinctively consumes more salt in a hot dry country. It also explains the importance of salt in Africa, the Middle East, etc., where caravans travel up to 1000 km to fetch salt; and at Taoudeni, the only town in the central Sahara, the monetary unit is a block of salt. Remember, too, the importance given to salt in the Bible, and that in the Middle Ages the vital role of salt was exploited as a basis for the universal (and unpopular) salt-taxes.

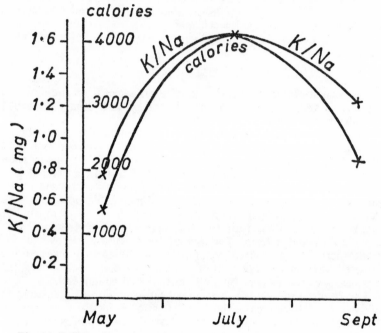

Fig. 14. Relation between K/Na rate and thermic rate in Sahara desert.

The passage of sodium to potassium was confirmed by a study in a still drier area of the Sahara, carried out with the cooperation of the French navy and lasting eight months. Systematic research was also carried out in a physiology laboratory, and all this work in dry heat has been explained in my publications. Here I give only the conclusions: physiological laboratory—a study by Lehmann—a man exerting a great physical effort for 3 hours at 39°C—humidity 60% —his urine shows a three-fold increase of the potassium content in relation to the sodium content—ratio gram-equivalent K/gram-equivalent Na.

This reaction is of great biological importance. It was known that persons suffering from a lesion of the adrenal gland excrete large quantities of potassium even if none is administered. It was never understood where the potassium originated; the low disposable reserve in the body did not justify such a massive excretion. On the other hand, it had been observed that in certain types of illness (for example Addison's disease) the salt in the body tended to decrease.

Blood plasma is a fluid especially rich in sodium chloride (sea salt), with an average content of 7 g per litre, but this salt content can decrease even when food is normally salted. This decrease in sodium was not linked with the appearance of potassium; it was treated as one of the mysterious and unexplained phenomena of life. Doctors had seen that potassium in the blood could increase to a dangerous level. This led to a blocking of nerve excitation accompanied by appearance of an equal electrical potential on the two faces of the nerve cell wall. Normally the external cell face has more sodium and less potassium than the internal face; and these two elements have different electrical potentials. It is the ratio between the external and internal potassium ions which controls the " membrane potential ", or the difference in potential between the two walls. (More exactly this potential is proportional to the logarithm of this ratio which is reduced to zero when there is the same number of K ions on each side of the membrane). Then (even a little sooner as there is a threshold) comes a stage where the nerves of the heart and lungs are annihilated and syncope and death follow. Some doctors have thought that by withdrawing plasma too rich in potassium and replacing it with an artificial serum containing only sodium chloride they would obtain a remission. Unfortunately, at every attempt, death immediately occurred. A prominent professor of the Faculty of Medicine in Paris asked me to go to his department of endocrinology in one of the largest Paris hospitals to explain to his students, assistants and collaborators exactly what took place.

The reader will have understood from what has previously been disclosed that potassium comes from sodium. If a new supply of sodium is injected into the organism, it is rapidly transmuted into potassium.

The substance which gives rise to this transmutation was found to be aldosterone. In cases of diseased adrenal glands the secretion of the opposing hormone is insufficient and the equilibrium is upset. A similar situation exists in cases of incipient hyperthermia; under the influence of the pituitary gland an accelerated production of aldosterone occurs, which will transmute available sodium into potassium by a thermo-energetic reaction. The body thus has a previously unsuspected thermo-regulator. The process by which aldosterone is secreted is complex, and is outlined in *Biological Transmutations*.

It should be realised that reactions of this type are governed by the physiological conditions of the subject. The number of calories in foodstuffs as calculated in dietetics has only a relative value, being based on the chemical energy liberated by combustion of carbon in the foods, especially the carbohydrates (sugar), whose values have been determined by chemical experiments in the laboratory.

53

The sodium-potassium bond exists in many forms. For instance, in a study of land and marine iguanas, it was shown that some species secreted, by a specialised nasal gland, a fluid containing up to 190 times more potassium than the blood plasma, at a rate of 190 cc per hour. An injection of a sodium chloride solution into the cloaca of these reptiles increased the potassium in the nasal secretion, but there was no increase of sodium; whereas with an injection of potassium chloride, there was a simultaneous increase in volume of secretion and of its potassium concentration. It is now understood that these nasal glands form a part of the heat regulatory system. When lying on a rock in the tropical sun the skin of these iguanas may reach a temperature of 70°C, but there is no evidence of their blood being at the same temperature; this remains to be verified. (In man the normal internal temperature is 0.5°C higher than the rectal. This difference can increase to 1° or 2° before triggering off the mechanism of endothermic potassium production, such as occurs when working in high temperatures or in cases of fever. In rare instances the temperature may reach 40°C before isothermia takes place at this higher level.) It was mainly research on humans which led me to think that transmutations take place, and then to confirm them. I could do this in the course of my duties, since I was able to make use of official laboratories for analyses relative to carbon monoxide, and to work in dry heat, etc. I thus had the opportunity to establish the essential principles of transmutations; but the verifications made with laboratory animals, and with others (chicken, crustacea, etc.) were carried out privately. The role of certain micro-organisms was ascertained in the same way, and would have been impossible by chemical process.

RESEARCH ON FISH

If certain fresh water fish are put into a tank of salt water, they can survive for several hours, even several days, depending on the concentration of salt.

Here are the results of a test made with a tench in water containing 0.8% salt. Figures are grams per litre of total blood:

	NaCl	KCl	CaCl₂
Before placing in salt water	5.10	3.95	0.30
After 72 hours	6.40	5.39	0.30

In water containing 1.4% salt, the figures are:

	NaCl	KCl	CaCl₂
After 4 hours	6.80	5.40	0.30

(Analyses made by A. Jullien and co-workers, 1959—Faculty of Sciences, Besançon.)

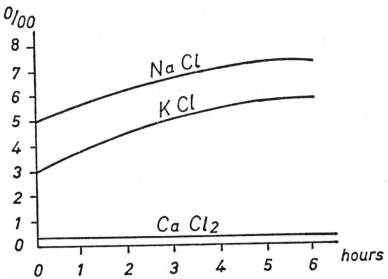

Fig. 15. Na and K variations in a freshwater fish placed in salt water.

Therefore there is no change in the rate of calcium; there is a small increase of sodium; and the potassium in the total blood supply (plasma and corpuscles) increases by 60% with the water containing in addition only sodium chloride. There could have been no movement of potassium from the blood corpuscles to the plasma, but to meet the objection that such a movement takes place, analyses were made of the total blood. Moreover it could not be said that Na and K increase was the result of a loss of tissue fluids, because in that case there would have been the same variations in concentration of all the salts; here calcium had not changed at all.

*
* *

Potassium metabolism has always been of interest to biologists.

Woton has shown that potassium continues to be secreted by the kidneys for several weeks after adoption of a very low potassium diet.

Lehmann, director of the Physiological Laboratory of Dortmund, states: " The quantity of potassium excreted is no indication of the quantity absorbed." He also has written: " The increase in potassium excretion during work in high temperatures cannot be attributed to a higher potassium intake." But the reason for it had not been suspected.

A number of laboratory studies on organs cultured in artificial media has shown (see *Biological Transmutations*) that with an increase in potassium concentration, the metabolism in the cultured tissue takes place at a higher temperature. This would explain the " reflex " action of the organism in order to maintain vital functions in conditions of elevated temperature.

(2) THE SODIUM-MAGNESIUM BOND

Here are the results of the investigations made in the Sahara. They confirm the reaction in which we have:

$$\underset{11}{\overset{23}{Na}} + \underset{1}{\overset{1}{H}} := \underset{12}{\overset{24}{Mg}}$$

(Sodium + hydrogen = magnesium)

Also here are figures in mg per man per day:

	Ingested	Excreted	Difference
April	288	290	− 2
May	247	354	− 107
July	348	528	− 180
September 5-9	198	420	− 222
September 12-16	211	286	− 75
Average	258.4	375.6	− 117.2

Therefore an average of 117 mg more metallic magnesium was excreted than ingested daily, but there was a decrease in sodium. So the sodium must have changed into magnesium.

This fact was apparently so abnormal, that three years later a further investigation to confirm the results was made by a new team of doctors and new analytical laboratories. The experiments were carried out in a still drier area and extended over eight months.

I will give only the daily average per man for the eight months:

Ingested	Excreted	Excess
314	570	256

The excess of excretion over ingestion was 256 mg for the average of the tested periods; this represents an increase of over 80%.

In April the shade temperature in the Sahara is sometimes lower than that of the human body, and there is then less excretion. It should be understood that that part of the body away from the sun is therefore in a surrounding air temperature of less than 37°C, and as a result this part of the body can lose heat.

However the doctor responsible for the tests was surprised by excess of excretion over ingestion. He suspected a magnesium deficiency, a " hypomagnesemia ", and increased the magnesium intake in drinking water (which had only a small amount of magnesia) and food.

56

To his great surprise, doubling the quantity absorbed had also doubled the excess excreted over ingested. Here is a comparison between a hot month and one of lower heat.

	Ingested	Excreted	Difference
April	220	530	− 310
August	395	1047.5	− 652

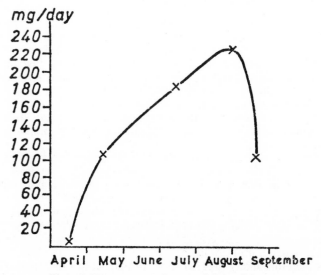

Fig. 16. Mg rate variations in Sahara desert.

A calculation, taking a daily average of 256 mg over eight months would show 61.44 g magnesium excreted by the body, while only about 5 g were available (there are also about 20 g in the bones but in a fixed state). In one week in summer these men should have lost all their available magnesium; in fact they lasted out for eight months.

These various experiments plainly confirmed the " manufacture " of magnesium by the organism under working conditions at high temperatures. (In low temperatures the process is reversed; the body must be supplied with magnesium.)

We have thus shown that sodium, with hydrogen, gives magnesium, and that with oxygen it gives potassium.

This explains that micro-organisms (including bacteria, though a complete study has not been made) are able to transform pockets of salt in the soil into potassium, and we believe there can be no other origin for the so-called potassium deposits, which in the crude state contain sodium chloride (not yet transmuted) mixed with potassium and magnesium salts. Where the deposit occurs within a more recent geological formation, it has a higher proportion of non-transformed sodium.

Sylvinite, the most recent deposit, has an average composition of 55% sodium salts, 25% magnesium salts, 20% potassium salts.

Of the older formations, kainite has no more than 35% sodium chloride, and carnalite no more than 23.2%.

So it is seen that more than half the sodium chloride in sylvinite has not been transmuted. This form of transmutation in the soil is very slow, as a salt pocket is practically devoid of water (the salt would otherwise dissolve away); the bacteria would have little hydrogen to give magnesium, or oxygen to give potassium; nor would the temperature favour their proliferation. One wonders if it would not be in the interests of potassium mining companies in Alsace, Germany and elsewhere to " incubate " the extracted raw material, if necessary with added micro-organisms, at temperatures around 30°C in order to transmute the sodium into potassium rapidly; potassium alone is wanted. One salt bacterium proliferates even at 45°C. We have already seen that through the activity of certain fungi (the ascomycetes, moulds and yeasts) potassium can be obtained from sodium within three days.

The reason why these minerals must be treated at a temperature of at least 30°C is as follows: we have seen (in *Biological Transmutations*) that with *in vitro* cultures the higher the temperature the more potassium is required; potassium is associated with heat; conversely magnesium is associated with cold. (Remember that in dry heat of at least 38°C potassium increases in the human body, and that living tissues stand cold better when the culture medium is enriched in magnesium.)

It is therefore likely that, when more magnesium than potassium is produced in the deposits, the temperature is too low, resulting in a predominance of magnesium-producing microorganisms.

Bulk storage at a minimum of 30°C would perhaps suffice to increase the potash content in crude potash. It might also be that with the most active bacteria the optimal proliferation temperature could be 40 ± 2°C.

In certain deposits the predominating mineral is magnesium. Crude kainite contains 29% magnesium salts, but only 20% of potassium salts. Carnalite has 32.4% magnesium salts, 16% potassium salts and only 23.2% sodium salts.

This takes us back to the most common reaction:

$$\text{Sodium} + \text{Hydrogen} \quad := \quad \text{Magnesium}$$

although the most useful but less common is:

$$\text{Sodium} + \text{Oxygen} \quad := \quad \text{Potassium}$$

The latter reaction is produced by " oxidising " micro-organisms, and it would be necessary to isolate and cultivate them and add them to the sodium chloride in order to change it into potassium chloride. Again, a simple study of optimal conditions for selective proliferation would permit enrichment of crude potash—at least, this would be possible if the conditions differed from those for proliferation of magnesium-producing organisms. From observations already noted, this seems quite feasible, though conditions for magnesium production by micro-organisms can differ from those in tissue cultures. We have seen the high magnesium production by man at high temperatures, while at low temperatures magnesium must be supplied. Precise extrapolation and *a priori* pronouncement cannot be made in biology, systematic investigation must be made, but to date it has been impossible to initiate it in the laboratory.

It should be remembered that when lakes are formed by closure of bays, thereby cutting them off from the sea, (or by the raising of the ground—Scandinavia and Canada) the relative concentrations of the various salts are altered. These changes are therefore due not to evaporation, but to activities of bacteria. Salt deposits, which are residues of ancient seas, may have an increased mineral concentration, especially of magnesium, through this bacterial effect. In central Sahara, the salt deposit at Taoudeni is much richer in magnesium than sea salt, and most of the water in the Sahara is of the magnesia type.

(3) THE POTASSIUM-CALCIUM BOND AND THE MAGNESIUM-CALCIUM BOND

An observation which had interested me, and contributed to my scientific " re-alignment ", was that concerning the presence of saltpetre on limestone walls. The fact had long been known, and saltpetre was used to make gunpowder long before it was known, in the last quarter of the nineteenth century, how to manufacture saltpetre (which is potassium nitrate) from potassium chloride. Saltpetre was not only scraped from walls, it was also swept from the soil in limestone districts where damp and warm periods alternated with dry ones. At the beginning of a dry period the soil was covered with a thick white layer resembling snow, and after a first sweeping, a second could generally be made a little later.

On a wall of a house I had on the coast, I noticed a continual " growth " of saltpetre despite repeated scrapings. Certainly the limestone itself was not the source of the quantities of potassium which I scraped off as saltpetre, several times a year, for eleven years. I thought, therefore, that the potassium could only have come from the calcium; and calcium − hydrogen = potassium:

$$\underset{20}{\overset{40}{Ca}} \quad - \quad \underset{1}{\overset{1}{H}} \quad := : \quad \underset{19}{\overset{39}{K}}$$

I compared this with the case of the hens which, with no available calcium, lay eggs with a calcareous shell when they have access to mica, which contains potassium silicate. The hens would use the opposite reaction: potassium + hydrogen = calcium.

59

So here is a reaction which nature has made reversible. The similarity of the two reactions was one of the facts which led me to the
hypothesis that transmutations take place by addition or subtraction
of hydrogen by means of displacement of a single proton from a
nucleus. This opened the way for further research using a single
proton as a variable.

It should be noted that crude saltpetre is a mixture of calcium
nitrate (not transmuted), potassium and magnesium nitrates (in industrial saltpetre production only potassium nitrate is left after dis-
solving out and evaporating). On the nuclear scale, in passing from
calcium to magnesium, we have:

$$\overset{40}{\underset{20}{Ca}} \quad - \quad \overset{24}{\underset{12}{Mg}} \quad := \quad \overset{16}{\underset{8}{O}}$$

In this way, bacteria produce magnesium by taking away oxygen
from within the calcium.

This immediately explained the " metasomatosis " of dolomite
rocks; the magnesium of this carbonate coming from the calcium
carbonate within which it forms.

Many phenomena met in agriculture can also be explained. Magnesium is present in plants and is, thus, an element indispensable to
them (the chlorophyll molecule is built up around a nucleus of magnesium). Even when no magnesium fertilisers are applied, many plants
take 20 to 50 kg of magnesium out of a hectare of soil annually. A
good virgin arable soil contains between 30 and 120 kg of magnesium
(soluble and insoluble) per hectare, so that after about two years of
cultivation such a soil should become exhausted. One author con-
cludes " the majority of arable fields should be very quickly exhausted, but in practice this is not the case."

But if the magnesium is not supplied to the soil, where does it come
from? It was not thought to have come from lime dressings, nor
from that naturally present in the soil; but it had been noticed that
when more lime was supplied to plants, their magnesium content increased. The explanation for this has been given above.

Let us now look at some research which confirms the bond between
calcium, potassium and magnesium.

Hens The hypothesis I had formulated on transmutation of potassium to calcium required proof. Some hens were put in a chicken
run with a hard clay soil, and left without lime; after a few days they
had used all their calcium reserve and laid soft-shelled eggs. That
same day they were given a supply of pure analysed mica supplied
by La Société des Kaolins d'Arvor de Lorient. This company supplied
70% of the national Kaolin production in 1965. Mica is a constituent
of granite; the felspar in granite changes to kaolin; silica and mica are
eliminated.

The hens, purchased as day old chicks, had been kept in an enclosed area and had never seen any mica, but they fell upon it with distinct relish, and pecked so quickly that they became breathless. They stopped for a while then began pecking again. Instinctively they had recognised a substance which would correct their deficiency. The following day the eggs had normal shells.

The experiment was repeated and extended to 40 days: whenever the mica was withheld, soft-shelled eggs reappeared. This confirmed what was already known: potassium cannot be stored up, and is both rapidly formed and eliminated.

The supply of potassium was transformed by the hens into calcium within 20 hours. An egg takes 8 to 10 days to be formed; there is a chain of eggs inside the hen, and the eggs become larger towards the point of laying. Only one egg at a time has a shell, and as one egg is laid the next starts to form its shell; the composition of the egg itself depends on the food eaten a week before it is laid, but the shell is produced by a rapid secretion.

CRUSTACEA

Shell formation in crustacea was a mystery—in fact, several mysteries. It was said, quite definitely, that the animal " fixed " the lime in sea-water, always the same explanation without proof but specialists who studied the phenomenon remained perplexed.

One day my young grandson brought me a crab which had cast off its shell for it had just moulted. It was a soft mass, and to keep it alive it was put in a basin of salt water. The following day it already had a firmer shell which was completed a day later. In thirty hours, a crab can form its shell which, if approximately 17 x 10 cm will weigh about 350 g. This total weight of fresh shell includes claws, partitions and water in joints). The amount of lime in sea-water is quite low, on average 0.042%Ca. The crustacean renewing its shell cannot obtain lime from other marine animals, in its vulnerable state it hides and does not search for food. Analysis of the animal's body has shown that only the hepatopancreas stores a little lime before the moult, but its shell contains 40 times more lime than this organ— Well then?

We have seen that magnesium (and potassium) present in sea-water (0.5% magnesium and 0.05% potassium salts) can give calcium, and it is essentially magnesium which is utilised by crustacea for making their shells. (We neglect Si because sea water contains little: 0.0001%).

At the Marine Biological Laboratory at Roscoff, a sea crayfish was placed in a tank of sea-water from which all lime had been removed by precipitation, and the animal formed its shell just the same.

The study of histological cuts from animals during shell secretion has shown that lime is formed on the external surface of a membrane where chemical analysis finds Ca. On the other, internal surface, the side for entry of matter, there is no lime. This fact has puzzled specialists.

The same situation exists in the formation of bone. Honest specialists accepted that bone formation (or shell in crustaceans) was an enigma. They only looked for lime, and didn't think to look for other things.

SALTPETRE

We have seen that a study of saltpetre was one of the subjects which led me to consider the possibility of transmutations of calcium into potassium (and magnesium). Nitrates form, therefore nitrifying bacteria are involved. These bacteria are associated with the micro-organisms providing potassium (by taking away hydrogen) and magnesium (by taking away oxygen) inside the calcium atom.

A laboratory investigation was made on a nitrogenous nutrient solution (ammonium salt) containing a mixture of bacteria. In the control flasks, the bacteria were killed by heating in an autoclave; thus the conditions were identical in all the flasks, except in half the flasks —marked—bacterial action was impossible. All the flasks were sterilised before inoculation of bacteria to destroy stray germs present in the water or in the nutrients or in the flasks or in the lime, the latter added and analysed before the experiment.

All the flasks were incubated at 28°C for 21 days. Potassium readings were then made on the contents of those inoculated with bacteria and of the controls. Here are the relevant figures, in mg of potassium per litre:

	Before Incubation	After Incubation	Increase of K
With calcium carbonate (chalk)	2.56	2.63	2.73%
With lithothamnus (marine algae)[1]	3.50	3.70	5.71%

There is therefore a significant increase in potassium.

Another experiment was made in which K and Ca were quantitatively determined in 15 flasks, showing an increase of 6% for K/Ca.

The experiment was repeated with as many as 100 flasks, and the increase in potassium was confirmed. Analyses were made of contents of the control flasks to ascertain the amounts of potassium coming from impurities in the material, in the water and from ion-exchanges with the glass (there is no absolute purity). When an ammonium salt is present, and especially with autoclaving, these ion-exchanges are not negligible. (There is always some potassium in glassware, though less in Pyrex, but even in plastic materials, polythene, styrene, etc. some potassium is present.)

[1] (Also called lithothamnion—*Lithothamnium calcareum*.)

(4) LIME AND SILICA

We have already dealt with the passage of silicon to calcium. We can say that nature has three means by which calcium is obtained: potassium + hydrogen; magnesium + oxygen; silicon + carbon.

Therefore, in this type of transmutation, the nuclei of hydrogen, oxygen and carbon respectively are displaced, and it is a curious fact that these three elements are also present in organic chemistry. Indeed, chemistry has only revealed a part of the organic phenomenon.

The silicon-calcium bond is evident also in detailed study of the incubated egg. The new-born chick has a bony skeleton which necessarily contains lime, but there is little calcium in the egg, and at birth the chick's skeleton has four times more lime than is present in the combined yolk and white of the egg.

It had been said that this lime came from the shell (always this same type of facile statement), but some research workers were interested in the disproportion between calcium in the skeleton and that in the egg, and wished to establish whether, indeed, there was a transfer of calcium from the shell. They were unable to prove it.

The same is true of reptile eggs. The young reptile on leaving the egg has a skeleton containing more lime than there was in the egg. The parchment-like shell ruptures on hatching and contains only a very small amount of lime (i.e. calcium carbonate; aragonite for tortoises, calcite for all other reptiles—including dinosaurs—as for birds).

It should also be mentioned that birds' eggs have essentially the same composition as those of batrachians and fish, but the two latter have no calcareous envelope. When hatched, however, the young have a calcareous skeleton even in soft water of districts without limestone.

Research scientists have noticed (see my book *Low Energy Transmutations*) that the weight of lime in the egg remains constant up to the tenth day of incubation. Then, the internal membrane becomes detached from the shell, the air space increases, and thereafter there can be no transfer of lime to the egg. This internal shell membrane contains organic silica, almost 0.5% in its outside layer compared to the weight of fresh matter.

The lime in the egg continues to increase, averaging 0.04 g on the tenth day, 0.05 g on the fifteenth, and 0.06 g on the sixteenth. Then, ossification of the skeleton rapidly takes place; the lime increases to 0.10 g on the seventeenth day, 0.13 g on eighteenth, 0.17 g on nineteenth and 0.18 g on twentieth. Thus, the amount of lime increases threefold between the sixteenth and the twentieth days.

In the above-mentioned book I devoted a chapter to a detailed discussion of this phenomenon which has been a fascinating problem for the last 150 years. (See Prout's work in England in 1822) (See Fig. 17)

The most recent research into shell formation shows that the lime of the shell membrane " grows " with external contact, crystals appear

63

at scattered points, then join to form a solid curved layer round the shell membrane. But although the shell originates from within, the lime forms outside the shell membrane, there is no lime on the internal surface which is in contact with the organic matter in the egg. On the other hand, this membrane progressively increases its silica content from inside to outside. Analyses made by A. Charnot give 154.79 mg of silica (SiO_2) per 100 g fresh matter in the internal layer, and 464.80 mg in the external layer. So, once again, we have silicon and carbon " growing " from the interior to the exterior where it emerges as calcium. (It has not been established that calcite crystals could be " fed " by the oviduct, and more information is needed.)

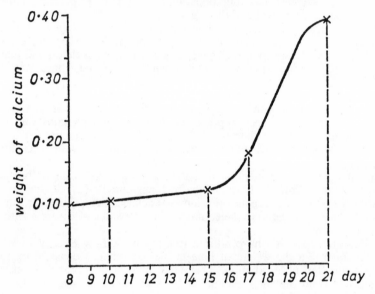

Fig. 17. Variation in the calcium rate inside a hen's egg during incubation (no change between the first and eighth day). (According to Prout's analyses). Compare with the graph according to analyses of Delezenne and Fourneau in *Transmutations à faible énergie* ". (*Low Energy Transmutations*).

From time to time I come across rather interesting quotations. We have all read texts relating to transmutations, but have not understood their meaning. When reading, facts can escape the eye if we are not already " tuned in ". In this way generations of people have missed the significance of this quotation from Flaubert, in *Bouvard and Pécuchet* (page 82, edition Garnier):

" Vauquelin, having calculated all the lime in oats fed to a hen, found still more in the shells of its eggs. Therefore, there is a creation of matter. In what way, no one knows."

Louis-Nicholas Vauquelin was a celebrated French chemist who died in 1829 at the age of 66, and was therefore a contemporary of Prout, and of Lavoisier who died at the guillotine in 1794. It can be seen that Vauquelin did not wholly accept Lavoisier's law. I have no other information on Vauquelin's research on the origin of lime in hens, but in subjects analogous to those investigated by Prout he must be credited with taking the first steps. In fact it was in 1822, when Prout published his work, that Vauquelin retired. He was professor at the Ecole des Mines and at the École Polytechnique in 1795, at the Collège de France in 1801 and a member of the Institute (a summit French institution) and director of the School of Pharmacy in 1803. He was professor of chemistry at the Muséum (1804) and at the Faculty of Medicine (1809). He left sixty memoirs, but I have not traced the date of the one cited by Flaubert. It is certainly earlier than 1822.

The origin of calcium in the hen continues to arouse interest and gives rise to a variety of research. Diets adopted for commercial poultry production include lime (calcium carbonate from chalk or from crushed shells) to provide a hard shell for the eggs. But Doctor Horwitz of the Institute of Agricultural Research, Volcani, Israel, stated in 1965 that although this mineral lime may be partly utilised in shell formation, it is also fixed in the tissues, notably the appendix. The hen's appendix has an important role in antibody production and its calcification renders chickens more susceptible to disease. This unwanted lime also inhibits activity of the growth hormones.

I will not go into a detailed investigation of the energies liberated in the reactions dealt with in this chapter. The subject, which has been raised in other parts of the book, has not yet been sufficiently investigated for any positive conclusions to be given.

I have been able to evaluate the amount of energy required to change sodium into potassium, but the figures must be accepted with reserve because the calculation derives from experiments made on only one case of its type.

There are other numerical values which record these energy changes qualitatively, but the measurements were taken from ions and not from atoms. A quantitative interpretation is not possible. For sodium to pass to potassium, energy is required (the organism takes it from its internal body heat). Sodium, which has a potential (as Na^+ ion) of -2.714 volts—referring to the hydrogen electrode—is replaced by potassium with a potential of -2.924 v. Therefore the difference of potential is raised by 0.21 v which means that energy is supplied. All the reactions due to a " merging " with oxygen are of this type: Mg $= -2.38$ v; Ca $= -2.87$ v; difference of potential $= 0.49$ v. Values are those of L.-C. Vincent. The values set by R. Lautié give for Ca $= -2.76$ or a difference of potential $= 0.38$ v. For reactions with hydrogen the differences in potential are in the opposite direction.

We make no conclusions, because the basic factors are different, but some previous investigations (cited by Bérard and de Larebeyrette in *Revue Générale des Sciences*, June 1965) show Na to have an "excitant" action, and Mg to have a "depressant" one. Similarly K is an "excitant" and Ca a "depressant". In terms of biological transmutations, Na furnishes energy (–2.714 v) and the resulting Mg has less of it (–2.38 v). The same applies for Ca and K. Some research on K/Ca ratio is reported in Chapter 8, part 2.

The concepts of the principal transmutations studied in this chapter are illustrated by this simplified diagram (that in Fig. 11 is more detailed):

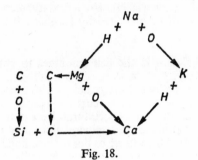

Fig. 18.

Chapter 7

RESEARCH ON PLANTS

> One must keep an open mind and admit
> that in nature, the absurd—according to our
> theories—may be possible.
>
> CLAUDE BERNARD

There has been a great deal of research on plants, but almost all laboratory work has been devoted to weight change in elements during germination.

A variety of reports reached me after I had begun to publish my results in 1959. Apparently several researchers had foreseen and had investigated these transmutations, but without being able to formulate any principles for such transmutations.

I will only mention these investigations briefly: they have already been cited in my other works.

In 1849 Vogel germinated cress seed in a medium recognised to be without sulphur, and found more sulphur in the shoots than in the seeds. This led him to state the hypothesis that sulphur might not be a simple body, since it could form itself. In fact transmutation may have taken place, but he did not say so.

Lawes and Gilbert made a series of investigations in England, from 1856 to 1873, and reported inexplicable variations of magnesium in the soil, and also a variation in weight of ash during germination of grass seeds. They reported their findings without comment or hypotheses.

Between 1875 and 1881 von Herzeele repeated Lawes' and Gilbert's experiments. He made comprehensive, exact laboratory tests in hydroponic cultures (cultures without soil) and found that a sulphate added to distilled water used for germination increases phosphorus in the shoots. (Note: sulphur $-$ hydrogen $=$ phosphorus.) He also showed after adding different potassium salts, that there is a rise in calcium. Von Herzeele drew up dozens of different kinds of experimental tables which remained forgotten because, at that time, what he had found could not be accepted, even if true. According to his contemporaries he was clearly mistaken and it was useless to repeat such " ridiculous " experiments.

However, in 1950 his fellow-countryman, Hauschka, published the forgotten papers and, in an appendix, pointed out the many aberrant phenomena in plant life which could not be explained. Spindler then introduced von Herzeele's works to France because, in 1946-47, he had already thought of the production of iodine by algae, especially Laminaria, a group he had studied.

Spindler informed Professor Baranger, head of the laboratory of organic chemistry at l'Ecole Polytechnic of Paris, of von Herzeele's research, and Baranger was interested in the results and conclusions. To verify them by modern methods he devised a technique using only double distilled water, and the same water with a small proportion of calcium chloride. He was able to show that germination in distilled water stops before the food reserve in the seed is used up—is it due to acid secretion by the radicles? Addition of a calcium salt maintains a pH suitable for growth.

Baranger's experimental results appeared in December 1960 in an English language journal published in Bombay. His detailed investigations demonstrated that phosphorus (in artificial light) decreases during germination of vetches. Von Herzeele had arrived at the same conclusion, but had found the opposite result for germ seeds in total darkness.

Another series of experiments, also on vetches, was carried out to ascertain variations in calcium, and showed an increase of calcium irrespective of external supply—or, at least, a greater increase than was justified by the amount available.

Professor Baranger did not then make simultaneous determinations of a number of elements, and did not discover, therefore, the relative increase and decrease of the elements involved. But one thing is certain, during seed germination one element can increase and another decrease. I realise that Professor Baranger has done much further research, but it has not yet (beginning of 1966) been published. He tells me that since his last publication in 1960 he has carried out, relative to the laws I have promulgated, experiments in which two elements were simultaneously determined. This has enabled me to show that iron can become manganese in ferruginous rocks through the activity of micro-organisms; Baranger demonstrated that germinating seeds induce the reverse reaction. Seeds contain a store of manganese: on germination this manganese disappears and there is an increase in iron due to the effect of an enzyme formed during germination. To date the enzyme has not been investigated and only the result has been confirmed. The reaction is:

$$\overset{55}{\underset{25}{Mn}} + \overset{1}{\underset{1}{H}} \; := \; \overset{56}{\underset{26}{Fe}}$$

In other experiments, Baranger made simultaneous determinations of a large number of elements. As for iodine production by Laminaria, which Spindler studied in 1946-47, this had already been the subject of lectures given by Freundler at the Sorbonne in 1925, and published in a small book in 1928.

I have the impression that Freundler envisaged that iodine might come from the tin in the granite on which algae are growing. He did not positively say so, but it seems indicated by his research on variations of iodine and tin occurring in the reverse direction.

Like von Herzeele fifty years earlier, Freundler came close to explaining the existence of transmutations. Both failed to do so. In von Herzeele's time, knowledge of the atom was exceedingly limited, and, it was not possible to establish a theory of transference of one element to another which might form a basis for research. Even in 1925 to 1928, Freundler lacked an important concept needed in order to visualise the transfer of tin to iodine: the concept of the neutron, discovered in 1932. He imagined a molecule formed of one tin atom —50 protons—and three hydrogen atoms, SnH_3, which would give a molecule with 53 protons corresponding to iodine. We can see that according to our theories this is impossible: the neutrons, and therefore the nucleons, were not accounted for; and at that time the nucleus was considered to be composed only of protons.

This question has an important theoretical interest. One often hears that algae " fix " the iodine of sea-water, but the reverse actually takes place: algae " manufacture " iodine. Each year, when the fronds (a little analogous to leaves) die, and every third year, when the stipes (analogous to stems) and the clasper die, the Laminaria liberate their iodine into the sea. (Contrary to earlier belief, stipes or claspers are not simply a means of fixation; they also act as roots and enable the plant to absorb tin from the rock; the tin passes into the stipes and subsequently into the fronds.)

A. de Cayeux, professor of geology at the Sorbonne, states in one of his books that there is no trace of iodine in undisturbed Precambrian strata, and asks himself where the earth's iodine came from as it was not present in former geological eras. We now know the answer. The iodine was " manufactured " at a later period, beginning in the Primary era with the appearance of large granite formations. Granite is a tin-bearing rock located within schists which, moreover, do not contain tin (the subjects of granite and iodine are dealt with in *Natural Transmutations*).[1] It should be noted that kelp (*fucus*), growing on schists and sandstone rock, " manufacture " bromine which, like iodine, is a halogen (bromine is in the seventh column of Mendeléev's table; fluorine heads the column, followed by chlorine).

As an addition to these investigations on plants, I would mention that the microscopic alga, Chlorellae, are powerful transmuting agents due to their intense enzymatic activity (they are able to produce thirty to forty times their own weight of dry matter in a few days). They can produce potassium when sodium is present, and research is in progress into transmutations in which they are involved.

---

[1] See also: G. Choubert—Publication UNESCO, in bibliography.

Chapter 8

APPLICATIONS AND DEVELOPMENT

> Our greatest progress in science is perhaps
> in the hands of men who are inclined and apt
> to ignore the artificial classifications we have
> set up . . .
>
> MELVIN CALVIN.
> Nobel Prize in Chemistry, 1961

(1) GEOLOGY

I propose to ignore those semi-industrialised applications which are now utilised to "enrich" minerals (copper, etc.) by biological methods.

In geology the origin of metals has always been a mystery. It is often said that they come from the "magma" of the earth's core and that they are found in "cracks" but, unfortunately for those theorists, ore is always near the earth's surface (even depths of 1000 or 2000 or 3000 metres are relatively close to the surface, since the earth has a diameter of 6000 kilometres). The quantity of ore decreases with depth and finally disappears; the mineral "fissures" or "chimneys", are really no more than small superficial cracks.

The eminent geologist Jean Lombard wrote in the preface to *Natural Transmutations*, 1963: "in order to stick to the rules of the geological game (which is very much a part of classical chemistry and thermodynamics) geologists have long relied on more and more complicated hypotheses which became more and more improbable until they were engulfed and hopelessly confused. They have supported these hypotheses to the point of losing cognisance of them."

Here is what he says of my work: "L. Kervran has opened up a wide field. The number of phenomena he includes as possible transmutations is such that it may be asked if the present confusion can hold out against him for very long. I should like to think not."

Professor Furon of the Faculty of Sciences in Paris, wrote the following when reviewing my book *Low Energy Transmutations* in *Revue Générale des Sciences* (February, 1965): "We know that M. Kervran has already published astonishing results from experiments showing that transmutation of elements take place in animals and plants, even in minerals. This was a fundamental discovery, obviously demanding new observations and new proofs. The work of M. Kervran has had a world-wide repercussion."

"This third book completes the two previous ones. It can no longer be denied that nature makes magnesium out of calcium (in some cases the reverse takes place); that potassium can come from sodium; and that carbon monoxide poisoning can occur without inhalation of the gas. In some scientific aspects of soils and metals, bacteria assume an unforeseen importance. Some transmutations are biologically beneficial, others are highly dangerous and it becomes possible to counter them. The whole problem of deficiencies must be reassessed."

" Mr. Kervran's discovery brings to light a host of new data of interest to doctors and pharmacists, biologists, dieticians, agronomists and farmers, and to geologists and those concerned with mineral exploitation. It is therefore hoped that the greatest number of research workers in all fields should give their attention to these problems with the aim of arriving at an explanation which might contribute to a better understanding of the approach and concept of the various factors involved."

So the geologists are fully aware that these new paths can lead to new horizons.

A candidate for a French teaching diploma, university degree, selected for a thesis: " The formation of iron ore deposits in Lorraine" and included three pages in allusion to my work.

Every informed person knows that iron does not come from the earth's central core. There is no connection between the core and the mineral strata; but all the classical theories speak of " concentration ", of water-borne materials, of hydrothermal eruptions and of deposits. Even if all this is accepted, these theories presuppose the existence of iron accumulated in certain locations. Therefore the iron existed but where did it come from? The examinee who had read my books, offered the explanation I had advanced, and at the oral examination was marked " very good ". She had shown evidence of a new point of view: if there is a concentration, where did the material concentrated come from? Afterwards, the examiners asked her to stay behind and tell them more about the subject (the examining board consisted of professors from the Sorbonne, from the College de France and the Ecole Normale Superieure; they were not all familiar with my work).

This problem of iron is dealt with in my books, notably in *Transmutations à faible énergie (Low Energy Transmutations)*. The production of copper and manganese is also included (plus a few lines on vanadium), since these two metals are certainly biological in origin. We have both iron and copper in our blood; if one increases there is a decrease in the other. Iron is the breathing pigment of haemoglobin. In certain land and marine animals, copper is the respiratory pigment in their blood (gastropods, cephalopods, crustacea, etc.). In other organisms (ascities, tunicates, holothuria, sponges), there is vanadium, but its exact role is debatable.

Although this problem of metals may be of economic interest to only a few people, it has great biological importance on account of the trace elements. Agriculturists should read the chapter and part of the appendix devoted to manganese in *Transmutations à faible*

71

énergie. (*Low Energy Transmutations*). Deficiency of this element in plants and in the soil had seemed mysterious: I have shown how it is associated with iron (the reaction is with $\pm H$ and is reversible:

$$\begin{array}{ccc} 55 & 1 & 56 \\ \text{Mn} + & \text{H} := & \text{Fe}) \\ 25 & 1 & 26 \end{array}$$

and the iron can be changed to manganese by micro-organisms (this is the explanation for the " black disease " of rocks, in which iron, changing to manganese, causes large black surface areas). At other times manganese changes to iron. Manganese is present in seeds, and on germination this element disappears to become iron in the growing shoots, as has been demonstrated in numerous experiments. Professor Baranger himself reported that the enzyme synthesised at the start of germination is capable of transmuting manganese into iron, and in some leguminous plant seeds, transmutes 25 times the weight of manganese, (to achieve this, a definite amount of soluble manganese salt was added to the distilled water).

Farmers and agronomists should read the chapters I have cited, as well as the one on copper. Many phenomena can be explained by transmutations: Jean Lombard, in a preface to a book by François Derrey *La Terre Cette Inconnue* (*This Unknown Earth*) wrote: " Though there may still be some resistance to these new discoveries by persons of good faith, we should admit that, in nature, sodium can become potassium, iron can become manganese, etc., and we are not entitled to fix *a priori* limits to such changes."

Jerome Cardan wrote in *La Tribune des Nations,* in reference to my book *Transmutations à faible énergie*: " Hundreds of experiments in reputable laboratories undoubtedly demonstrate that transmutations of atomic nuclei occur in living matter. It may be impossible, but it seems to happen. Sodium changes to potassium, calcium to potassium and vice versa. In certain cases silicon plus carbon gives calcium."

" Nitrogen is transformed to carbon monoxide. All of which is quite contrary to all known natural laws; but the experiments exist, and I know of no serious refutation of them. Instead, the author of this book offers confirmation, precise references and a complete bibliography for each particular case."

So at last, faced by these established facts, there is support from every genuine scientist. Research is conclusive, and from it geologists have a better understanding of the origin of metals and the different geological periods.

The geologist de Cayeux, of the Sorbonne, admits that there were no salts in the earth during the Precambrian era. Mineral salt formation began with silica in the Triassic system at the beginning of the Secondary, just following the Primary era. In the Triassic period/ system clay predominated but there was a small amount of lime. The process continued until the Jurassic period when lime predominated (for example, in "natural cement" rock) i.e. Portland cement.

At the end of the Secondary era, silica in certain places became changed into lime; this was the Cretaceous era. Salts (gypsum, etc.) are not present in those underlying Primary layers which had remained unchanged, and de Cayeux asked where these salts came from, and where their components originated. Only some of the components making up the salts existed in the Primary era and none at all in the Precambrian. We have furnished the explanation.

The following figures illustrate the impossibility of the oceans being responsible for the changes leading to the Secondary era. The earth has beds of Triassic formations estimated at more than 5000 metres thick; in places mixed formations are more than 9000 metres thick, some dating from the end of the Primary and the beginning of the Secondary. In the Jurassic there are some, more than 3000 m, and Precambrian formations, as calculated by seismic methods, often exceed 15,000 m.

Limestone did not only originate from silica, but also from magnesia (and some limestone of the Primary is dolomitic).

Only a few examples of the new perspectives opened to geology can be given here. Many phenomena now have a more logical explanation than was once the case and which only satisfied the imagination of its author. If such explanations were accepted, it was because of the absence of any convincing alternatives. The examples given of the passage of manganese to iron and vice versa permits a better understanding of the composition of the mineral tungsten W—wolfram, or more particularly wolframite, which contains all intermediates between 100% iron and 0% manganese, or between 0% iron and 100% manganese in combination with the radical WO_4. The mineral containing only iron is ferberite, $FeWO_4$; that with only manganese is hubnerite, $MnWO_4$.

I have cited[1] Dr. Charruyer, professor at the School of Medicine in Limoges, who informed me of the presence of calcite geodes in very hard, compact, totally impermeable schists in the primitive strata west of the Massif Central.

In a similar context I recall a paper given by Daugatry, Perry and Williams, at a symposium on "The Geochemistry of Sedimentary Carbonate Rocks" at Copenhagen in 1960, dealing with anomalies in development of dolomite rocks. At the same symposium Seibold discussed isolated carbonate concretions found in the lower Jurassic without fissures and said: "The concretions are surrounded by dolomite, but have none of the metal inside. This comparative wealth of $MgCO_3$ is probably not due to an external source." His statement remained unexplained (but in Jurassic formation, lime is present and we now know that magnesium originates from it).

These are only a few examples but there are numerous instances where classical geology had reached a dead end.

[1] See *Transmutations Biologiques* (*Biological Transmutations*)

73

2 MEDICINE

Transmutation, this special property of matter finds many applications in medicine.

Potassium in the human body is derived mainly from sodium, and in many disorders there is an overproduction of potassium; in such disorders, therefore, the intake of sea salt should be reduced. On the other hand, since salt is required in order to produce magnesium, and magnesium yields calcium (as well as phosphorus), there could be a risk of mineral deficiency. In such cases the doctor is able to reinforce the amount of minerals in the body by giving absorbable magnesium in the form of its carbonate or chloride, or by giving organic silica. Magnesium and silicon can give calcium but not potassium.

Certain arthrosic conditions have been cured by administering organic silica and a small quantity of added potassium bicarbonate. And doubtless some forms of rheumatism could be cured in the same manner.

Atherosclerosis presents a similar situation. When an atheroma forms the arteries have usually become porous, as a result of loss of silica; porous calcium forms within the arterial walls and arteriosclerosis follows.

Such lime deposits occur widely (in walls of blood vessels and of the heart, in muscles, or in the form of calculi) and usually originate from the organic calcium ingested with the food. Inorganic calcium is not assimilated and is therefore excreted. By a process described in *Transmutations à faible énergie* (*Low Energy Transmutations*) organic silica is able to dispose of this parasitic lime. New techniques, therefore, are becoming available, but they have not been fully studied.

The reaction of potassium production can be used in treating disorders due to heat. During very hot weather, for example, drinks containing 10 grams of salt per litre may be taken (sea-water has approximately 29 grams sodium chloride per litre). In cases of fever the doctor will decide the quantity of salt to add to the liquid diet, which could be a salted broth, etc. To prevent hyperthermia, workers subjected to dry heat could drink flavoured or plain water containing 3 to 5 grams of salt per litre. At 5 grams there is a distinct salty taste; 3 grams is just discernible.

The problem of decalcification and of recalcification have to be reconsidered in the light of our new knowledge. It has been shown, for example, that healing of bone fractures can be very much accelerated by the use of organic silica (with a little magnesium and potassium).

Professor Delbet had already envisaged this in its negative aspect: " It is scarcely probable that calcium phosphate is formed in the bones." He also wrote: " We do not know in what form phosphates of calcium arrive in the skeleton ". In fact, calcium has never been found to enter into the bones.

Professor Stolkowski recognised, in a book devoted to the problem of lime in the body, that: " It is usual to hide our ignorance of the biochemical origin of calcium by giving the name of phosphocarbonated protein to a secretion of the formative cells." Thus, instead of an explanation, a word (the same is true in " metasomatosis " of " dolomite rock " and " soporific property " of opium).

In 1962, in Montreal, H. Selye wrote a substantial book on what he termed " calciphylaxis ", defined as " a state of diffuse hypersensibility in which the tissues react by an intense local calcification ". Such an explanation is purely dialectic. He concluded bitterly: " The nature of the local mechanism of calcification is one of the most important unsolved problems in " biochemistry ". He was not aware of my work.

This question of calcification also leads to a revision of our ideas on dietetics. No purpose is served by evaluating the quantity of mineral calcium in foodstuffs, our body rejects most of it and often only utilises the remainder badly (generally in hot weather more calcium is eliminated than is ingested). As calcium originates partly from magnesium, when the body " manufactures " too much magnesium it also " manufactures " too much calcium, and the calcium levels—the same as the phosphorus—are almost always related to those of magnesium.

Moreover, the concept of transmutation has passed unnoticed by conventional dieticians, which is why calorific values based solely on chemical reactions of carbon oxidation, are inadequate. We can no longer accept that a simple calorific reading is the measure of energy released in a chemical reaction. Elements can be transmuted by enzymes with a resulting absorption or liberation of energy, as in certain physiological conditions which either stimulate or inhibit enzyme production. All of which shows that the determination of energy values is a complex operation incompatible with the elementary technique hitherto practised. There were too many reservations and margins of error in it to be acceptable by enlightened scientists, whose own experiments had led them to doubt the accepted figures. Now, after an exhaustive and comparative study of calorific values we realise that the values determined by the old methods are false. In consequence, the science of dietetics is becoming still more complex, and problems of nutrition should everywhere receive more specialised attention. In Japan, institutes of nutritional science have been established in the universities. Most people have been unaware of the breaching of outmoded attitudes and rules which has come with the explosive development of scientific exchanges between one side of the world and the other[1].

I cannot go into the numerous applications of transmutations for medicine; in *Biological Transmutations* and in *Low Energy Transmutations* there are many observations concerning sodium, potassium,

[1] See my introduction to *Sauvez votre Santé* (*Save your Health*) by Dr. Plisnier, Editions P.I.C., Brussels, 1966

calcium and silicon relevant to the human body. In the latter book, reference is made to the bond between iron and copper which explains the disorders such as Wilson's disease, in which a continued increase of copper, leading to paralysis and death, could not be prevented. The bond between fluorine and phosphorus is also mentioned. Calcium phosphate (bone or mineral) is never present without fluorine

$$(\ _{9}^{19}F \ + \ _{6}^{12}C \ := \ _{15}^{31}P),$$

or rather $_{15}P - {}_{6}C := {}_{9}F$, which explains how F increases in fossilised bones, even among specimens in museums (microorganisms transfer C from P). Formerly this was an inexplicable anomaly, and a simple deduction leads one to wonder if dental caries are allied to this phenomenon. This would be worth studying since, in any case, the introduction of fluorine can result in phosphorus. In advanced cases of osseous demineralisation and of chronic articular rheumatism Dr. Charnot has obtained remarkable results by adding a very small amount of fluorine to organic silica and potassium. Silica and potassium then become calcium, while fluorine reinforces the phosphate (which also comes from magnesium). The problem of fluorination of drinking water has been raised in this connection. The practice has been adopted in some countries; in others it has been rejected because the quantity of flourine added to tap-water— to obtain any beneficial effect—would reach an injurious level where water and other liquids are consumed in large quantities. The matter is altogether different when fluorine is given under medical supervision, solely where necessary.

Bone consists of one third organic interstitial matter, and two thirds mineral salts forming a structure.

The average composition of bone mineral salts according to individual, age, state of health, is:

	per cent
Tricalcium phosphate	85
Magnesium phosphate	2
Calcium carbonate	9
Calcium fluoride	4

The percentage of fluorine is by no means negligible and should not be ignored.

The impossibility of a positive reliance on chemistry is evident to many medical practitioners. For example, where recalcification is required, a diet rich in calcium does not necessarily achieve it.

This particular aspect has long been recognised by those who have given their attention to the promotion of a correct diet. Doctor E. Plisnier, a Belgian specialist in modern dietetics, has made a number of statements explicable in terms of biological transmutations. In his book *Sauvez votre Santé* he makes several observations which will only be quoted in part here: "Children with retarded dentition receiving a normal amount of lime in the diet (by classical dietetic standards) along with fruit, vegetables, milk, cheese and meat, have had the retardation overcome within a few weeks when milk and cheese (considered good sources of assimilable calcium) have been omitted.

"The same diet, poor in calcium, has led to the quick formation of a callus in a fracture." Dr. Plisnier cites in particular a case of a person over sixty years old who had a fracture of the neck of the femur. The classical methods of treatment had failed to heal it, in spite of two operations and a diet rich in calcium. A specially formulated diet, poor in calcium, brought about a recovery.

Similar instances have been reported by Dr. Charnot who has published details of several hundred cases, including rheumatoid conditions. Both of the doctors mentioned had observed that ingested calcium does not induce recalcification.

Dr. M. Montassut[1] also reports investigations showing a connection between depression, asthenia, alkalosis, and hyperemotivity. He cites several experiments which demonstrate the importance of the K/Ca ratio, the widening of which is a feature of hyperemotivity. In the cases studied, there was little variation in Ca, and the increase in K/Ca originated from a hyperkalemia, with an increased K level in the blood plasma. On the other hand, the administration of magnesium narrowed the K/Ca ratio (we have seen that Mg can give Ca). Therefore the ratio is narrowed by increasing the denominator; but this had not been understood.

The prospects open to medicine for an understanding of phenomena, hitherto empirically established, can now be realised. One result has been the administration of Mg in order to obtain an increase in Ca; the psychoasthenia studied by Doctor Montassut can be better evaluated, and the resulting therapeutic measures easier to understand and to apply.

In almost every medical journal, there are authors writing about contradictions they have noted, mainly concerning sodium, potassium, magnesium and calcium. The journal previously mentioned, for March 10th 1965, has a study of decalcification in an immobilised person, saying: "There is a pronounced negative calcium balance; nevertheless, the amount of calcium on the bone surface increases and there is simultaneously a three- or four-fold increase in calcium loss."

Calcium administration is of no benefit. It is often followed by peri-articular calcification: "Frequently, a urinary lithiasis supervenes which complicates the situation still further." But the authors are unaware of biological transmutations and are unable to see the solution.

We must not accept, blindly, all we have been taught. We were

[1] See *Gazette Medicale de France*, 1-4, 1965, pp. 1329-39

told that there were only substitution phenomena; that if calcium appeared it was because it had been liberated elsewhere, for example by magnesium, which is also divalent. This is merely a conformist position satisfying the dogma of non-creation of matter. Those who maintain the substitution concept have never made comparative studies of total magnesium and total calcium before and after definite time periods. Some authors claim there is substitution since total Mg + Ca remains constant, but this hardly helps their theory, because relative constancy is due to the fact that Mg has decreased and Ca has increased; they, however, maintain that when an element has disappeared in an organ it has gone elsewhere where, obviously, it could not be determined, since it could not be found.

We have already seen instances in which substitution alone could not have been possible. Highly exact analyses on germinating seeds have shown that when magnesium decreases there is a calcium increase, and the experiment was such that the vanished magnesium could not have gone elsewhere, it had positively disappeared; nor was there any possible external source of calcium. Tables in *Low Energy Transmutations* show the relevant figures.

In *Natural Transmutations* I stated that Sahara oilfield workers excreted a daily average of 320 mg more calcium than they ingested, and no decalcification occurred. Therefore, some other ingested element must have enabled the body to make this calcium, which could not have come from chemical substitution, because the body's available reserves were insufficient for such a great calcium excretion (averages were for a six-month period).

Everything written on calcium metabolism which ignores transmutations must, therefore, be reconsidered; there are important medical aspects.

A few examples selected at random will show the extent to which medical knowledge must be reconsidered. The interpretation of some experiments must also be reassessed, as the following example shows: " Bacteria make radioactive mercury volatile."

In industrial medicine, examinations have to be made of persons working with, or preparing mercury. In order to prevent mercury poisoning, or hydrargyrism, the level of mercury, particularly in the urine, must be ascertained, and, to avoid errors, samples are often sent to different laboratories. For a long time inexplicable variations in results had been observed between different laboratories which could not be attributed to the analytical techniques employed.

It has been noticed that after the addition of toluene there is no loss of mercury; mercury was said to evaporate, and the toluene prevented this evaporation. To carry this study further, L. Magos with A. A. Tuffrey and T. W. Clarkson of the Medical Research Council Laboratories, Carshalton, Great Britain, experimented with crushed and homogenised rat kidneys with addition of $HgCl_2$ marked with radioactive ^{203}Hg isotope. Samples were measured for radioactivity by Geiger counter at staggered intervals. The half-life of this isotope is 46 days, but the decrease in activity (corresponding to the disintegration of the radioactive isotope) was much greater than that calculated for the life duration of radioactive mercury. Where had the mercury gone? Since it had disappeared, this apparently confirmed its evaporation as claimed in classical science, but the explanation is unsupported and dogmatic, based on the sole fact that in a figurative sense mercury "volatilises", and the term was accepted in its literal meaning.

However, by chance the investigation revealed that mercury did not "volatilise" at the rate predicted. In theory, there should have been the usual, regular exponential curve according to the law of radioactive activity reduction. A count made 16 hours after the first check clearly showed a decrease of mercury. But after 32 hours the extent of the decrease could not be understood; 48 hours later the gap between actual and calculated decrease had widened still further.

To their credit the researchers thought that the results could be due to bacterial activity, and that the latent period of at least 16 hours could correspond to the incubation period for a bacterial colony. Could it be that the action, explained later, of toluene destroyed the bacteria? *A priori*, this was a risky hypothesis since it was accepted that bacteria do not attack heavy metals, which resist all biological action.

But the investigation confirmed the presence of bacteria, which apart from one species, were isolated and identified. Here is part of the results reported by the authors:

Volatilisation of Mercury by Bacteria
British Journal of Industrial Medicine
October 1964, pp. 294-298

Incubation, in hours	$HgCl_2$ Concentration	Remaining activity of ^{203}Hg in per cent			
		0	16	32	48
Normal decrease		100	99	98	97
Penicillamine dialysate	10^{-4} to 5.10^{-6}	100	99	98	97
Homogenate	5.10^{-5}	100	99	67	47

Therefore, the addition of penicillin killed the bacteria and the disappearance of mercury followed the same course as a solution sterilised in the autoclave. On the other hand, 50% more mercury disappeared in the inoculated medium than could be attributed to the disintegration of the radioactive isotope.

If a sterile sample is inoculated with only 1.10^{-5} mole of $HgCl_2$ which has already begun to lose its mercury there is no latent period; proliferation is immediate. In 24 hours 60% of the initial activity has disappeared (this varied according to sample from 56%

79

to 63%), while a non-inoculated control loses only 2% activity.

The most active bacteria, those effecting a 30% disappearance of mercury in 48 hours, were identified as *Klebsiella aerogenes.* Another belonged to an unidentified species of the genus *Proteus;* another active microorganism could not be identified. There were a number of other microorganisms with little or no action which could not be identified. In the water supply where the suspension of macerated liver was prepared, *Pseudomonas pyocyanea* was found. This microorganism is very active, as is also a *Diplococcus,* found in the water, which is partially inhibited but not killed by toluene.

Comment: My impression of this series of experiments is that at no time had the investigators thought of anything other than a disappearance of mercury by volatilisation, but this they had not confirmed. They had accepted it as self-evident, based on the decrease in radioactivity measured by Geiger counter, and taking as point of departure the addition of a definite quantity of radioactive tracer. For their conclusions to be valid they should have recovered and analysed the mercury vapour.

Incontrovertible results have thus been given an arbitrary interpretation, lacking in validity, due to the absence of an essential measurement.

It did not occur to them that the disappearance of mercury could be due to a biological transmutation of mercury into something that might have been found in the residual solution. Had it been found, it would be evidence of transmutation. Evaporation is obviously very slight, mercury boils at 360°C and between 40 and 50°C has a very low vapour pressure—their experiments were conducted at ordinary room temperature (about 20°C?). A more critical approach might have prevented the researchers from accepting the postulate of evaporation. The experimenters also assumed that radioactive tracers behave biologically in a manner similar to stable isotopes. In fact this is not the case; it can be taken as approximately so in some cases only.

These experiments lead to a slight modification of our previous statements on radioactive elements. Previously there had been no evidence that higher forms of life (animal or plant) carry out biological transmutation of radioactive elements, as radioactivity destroys the more complex types of cells. But bacterial behaviour is full of surprises and, as we have said (in *Low Energy Transmutations*), some bacteria grow in pure sulphuric acid (in connection with the change of iron into copper). We have also mentioned certain *Pseudomonas* living in the heavy water at the heart of an atomic reactor and receiving many thousand times the lethal dose for human tissues. Similarly *Micrococcus radiodurans* resists 3,000 times the lethal dose for a mammalian cell. In experiments on mercury, a species of *Pseudomonas* was found which " digests " radioactive mercury (to transfer it into another element, as yet unidentified?); but then what does it do with the excess neutrons in the radioactive nucleus? Since radioactivity decreases, there can be no transmutation into another radio-active element. Could there possibly be "a conversion" of neutrons into protons? Only a close study of the result of the experiment can provide the answer.

Might this not be a new line of research into disposal of radio-active waste? I advanced one possibility as a hypothesis in 1960 (see also *Natural Transmutations*). The above mentioned experiments seem to show that this is being done. Enough information has now been acquired to establish a research programme.

The disturbances due to γ-rays are responsible for inactivation of thymine in the DNA which for this reason cannot reproduce unless the damage has been repaired by enzyme synthesis. A badly damaged DNA molecule cannot reproduce, and progressive death of the cells follows. But with bacteria the speed of reproduction, and therefore of recuperation, is generally quite considerable. This indicates a very high enzymic activity and also accounts for bacteria withstanding irradiation several thousand times greater than the lethal dose for higher forms of animal and plant life.

The most resistant species to γ-rays seems to be *Micrococcus radiodurans*. Up to doses of a little more than 1,000,000 r (rads) this organism can produce thymine in the DNA faster than γ-rays in-activation can destroy it. In this bacterium the enzyme is apparently undamaged by such doses, as also are the other DNA bases; it is a question of absorption capacity and there are organic compounds which are more sensitive to ultra-violet rays than to γ-rays. Nothing as yet explains how a bacterium is able to react against a radio-active atom; is it by "digestion" or by a form of antibody secretion? On this point the problem remains wide open.

As the lethal dose (killing 50% bacteria) is a little above 1,000,000 rads, the question arises of the theory of survival by adaptation to natural environment: we do not know any natural environment in which such irradiation intensities occur.

3 AGRICULTURE—AGRONOMY

Certain questions which formerly remained unanswered are now less obscure; for example, the potassium "cycle". Estimated figures (1955) put the potash intake by annual plants in France at 1,500,000 tons (metric) per year, expressed in K_2O. Potassium supplied by manure was 300,000 tons, and by potash fertilisers 450,000 tons, so only about half the potash taken up by the plants was available from applications due to cultivation. Therefore, ever since cultivation existed, 750,000 tons have been supplied by the same soils annually (the application of potash fertilisers is comparatively recent; in 1900 only 60,000 tons were used, so at the beginning of this century the deficit was apparently much greater than now). But as the soil does not hold any appreciable potash reserves, how can 750,000 tons be taken out every year?

It is now understood that plants have two methods of "manu-facturing" the potassium in their tissues (which has not been sup-plied from an external source): the reactions sodium + oxygen = potassium (sodium may be obtained from seaweed); and calcium − hydrogen = potassium.

It is known that sodium nitrate can be used as a fertiliser, but although potash is always found in the ash of plants, sodium does not accumulate in the soil (in France, sodium nitrate was once widely used; after 1920 when Alsace, a source of potash, was returned to France potash replaced sodium nitrate). But if the plant does not need nitrate, which can have a wilting effect, another salt must be found for the sodium. Trials should be done with cheap crude sodium carbonate, (Solvay soda) as opposed to the refined soda crystal which would be too expensive. Sodium chloride is injurious to plants on account of its chlorine, there is a special technique for its use when the soil will accept it.

I am not suggesting for one moment that application of potash can be dropped, whether supplied in manure, compost or in any other way.

In the experiments with moulds and yeasts, when potash was supplied crop yield and weight of dry matter were much higher than when plant cells had to make their own potassium (of course, in the absence of proper experiments one should not generalise and apply this statement to all plants). However, when potassium is supplied in cultivation, the potassium content of crops (relative to weight of dry matter) is twice as great as in crops grown without added pot-assium. I have insufficient data to say whether foods high or low in potassium are better nutritionally, though one can take a stand on the general situation. Certainly the traditions of the Far East, based on empirical observation, recommend avoidance of everything rich in potassium ("yin", and excess yin is harmful), and in many countries, France included, the high level of potassium in foods is widely criticised, and there is much con-troversy on the subject.

I can, however, refer to a recent study carried out at the Ohio State Experimental Station by J. Benton Jones (published in *Science*, 2-4, 1965, p. 94), which, if I interpret his results correctly, could lead to a new approach. Unacquainted with biological transmutations, he only reported his findings, and was unable to explain them.

Trace elements are indispensable. They enter into the structure of coenzymes, and if a metal is missing in the coenzyme, the enzyme itself is without effect. Among the important trace elements having a multiple role is molybdenum (Mo). After potassium deficiency, as revealed by leaves of hybrid maize, Benton found the leaves to have a much increased molybdenum content.

Here are average figures for leaves of different varieties (p.p.m. = parts per million):

Plants deficient in potassium		Normal plants	
Potassium %	Molybdenum p.p.m.	Potassium %	Molybdenum p.p.m.
0.97	2.0	2.70	0.5
0.56	4.0	2.49	0.9
0.52	2.8	1.73	0.7

Benton then made a systematic verification by varying the amount of potassium supplied to the soil. Here are the results in kilograms per hectare: (Fig. 17).

Potassium supplied	Degree of K deficiency	Composition of leaves	
		%K	Mo p.p.m.
0	severe	0.57	4
36 kg/ha	slight	1.79	1.2
72	none	2.28	0.9
108	none	2.49	0.9

There is no alteration in molybdenum content in the absence of a potassium deficiency. The reason for this negative correlation between K and Mo is still unknown.

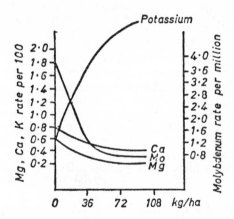

Fig. 19. Variations of K, Ca, Mg, and Mo in leaves of maize, as a result of potassium supplied to the soil. (According to Benton.)

I also reproduce a graph showing the simultaneous variations in potassium, calcium, magnesium and molybdenum. The author makes no comment on it, but readers can make their own deductions. The calcium decreases but the potassium increases and we have seen by what mechanism this takes place. Note that the calcium and magnesium are almost parallel on the graph. The reason is that the plant requires calcium to make magnesium for its chlorophyll; the more calcium is supplied, the more magnesium is produced in the plant (though a limit harmful to the plant may be reached). Conversely, if there is no longer sufficient calcium, there cannot be a sufficiency of magnesium. It is important to note that even if no potassium is given there is still some in the leaves, though four times less than if a rich supply were given. This situation is therefore analogous to that observed in laboratory cultures of microscopic fungi.

The above experiment is not comprehensive enough for any general conclusion to be reached. The origin of molybdenum cannot be deduced from it, but the particularly important piece of evidence is that a plant too rich in potassium is at the same time poor in molybdenum, a most important trace element. Consequently, one is led to ask whether research should not be made into the effects of excess potash on crop yields. The more potassium there is in plants, the greater the reduction of molybdenum; and what are the optimal quantities of these two elements in a plant? This does not appear to have been adequately studied, and there is not only one answer, since values differ not only between species but between varieties of the same species. This is a problem, however, for investigation by dieticians as well as scientists.

As for agriculturists, they know that excess potash, even if supplied by liquid manure, leads to a molybdenum deficiency. It is also known that a fruit disease termed "bitter-pit" in the U.S.A. is caused by too much K in relation to Ca. Similarly there is a Ca deficiency in tomatoes when there is too much K (see *Comptes Rendus de l'Academie des Sciences*, 1964, p. 3600). The K/Ca balance can be restored by supplying Mg, which then becomes Ca.

It has been explained in this book that if potassium fertilisers were no longer available it would not be a catastrophe for the farmers now using them. Equivalents could be obtained industrially or directly in the soil by at least two methods: yeast and microscopic algae could produce potassium from sodium; or other microorganisms (as with saltpetre) could produce it from calcium.

Perhaps there will soon be factories for growing microorganisms for agricultural and industrial purposes. Some are already in existence: yeasts are grown on an industrial scale, as are moulds for penicillin, actinomyces, streptomyces etc. but we should be thinking of extending such practices to include agricultural applications. Let me add that I have recently been told of a product developed by Professor Keilling, containing selected bacteria, that has just been put on sale.

French legislation on chemical fertilisers also requires modification as to nomenclature: potash is a fertiliser, lime is an improver, and we have seen that by bacterial (or enzymatic) means, the one can change into the other. Magnesium compounds are fertilisers, but dolomite rocks are not, etc.

The problems of mineral deficiencies in animals and plants must be reassessed[1]. In future it will be impossible for growers and stock raisers to ignore the phenomena of biological transmutations that they have all observed, and utilised, without understanding them. It is this lack of understanding which has limited their application.

This has been realised by the leading members of associations which advocate and apply biological cultivation in France, Italy, Germany, Switzerland, England etc. Such people are the pioneers of the agricultural community. They know that not all biology can be interpreted in chemical terms, and have seen the consequences of too much reliance on chemistry in a biological context. Excess, in anything, must be paid for sooner or later. (Advice is often given to apply 250 kg/ha of potash fertiliser containing 50% K_2O, or 625 kg/ha of sylvinite containing 18% K_2O. Cereals take up annually an average of only 40 kg/ha therefore two thirds of the K_2O supplied are wasted.)

The problems raised by transmutations are complex, and can have contrasting effects between species, and also in specimens of the same plant. For example, a plant with calcium available to its roots can transform it into magnesium. The seeds store a reserve of magnesium, calcium is required for the initial reactions, then, in the seedling, the magnesium reserve disappears and changes to calcium. This can only happen in sunlight; in artificial light, devoid of ultra-violet rays, the reaction does not occur. It is a reaction involving oxygen, Mg + O, and the displacement of an oxygen nucleus requires more energy than the displacement of a hydrogen nucleus. Reactions involving displacement of H can take place even in darkness. This is illustrated by addition of a sulphate to distilled water in seed germination where there is an increase in weight of phosphorus ($S - H := : P$). Similarly the reaction changing potassium to calcium can occur in artificial light ($K + H := : Ca$).

In every case one must see the significance of the reaction, according to plant species and the physical and chemical conditions (temperature, pH), the presence of specific micro-organisms, etc.—in other words, a completely new branch of agronomy has to be created.

Henceforth, agriculturists have proven data that cultivation is first and foremost a biological matter. All forms of cultivation conducted outside this concept must lead to a more or less rapid exhaustion of the soil.

Cultivation based on classical chemistry alone, fails wherever intensive and abusive methods are employed. The marked crop increases due to chemical cultivation last only a certain time, long enough to stimulate and over-feed microorganisms which produce enzymes at too high a rate. In spite of their astonishing ability to adapt themselves to most unusual conditions (we have given some examples in medicine), these organisms can finally be destroyed by prolongation

[1] In France, H. Rosenauer has developed a method of investigating such deficiencies by leaf analysis. It enables a better understanding of what the plant lacks. Soil analysis is inadequate for this purpose, it ignores the biological activities of the micro-organisms involved.

of conditions which prevent their normal reproduction. The soil then dies and remains sterile.

By such methods large areas in America have been lost to cultivation. In Western Europe the evil is so far less apparent. The farmer's common sense in the use of chemical and biological processes has largely avoided the abuses; but this has not happened everywhere, in some places deplorable results have followed. Warning signs are to be seen, inexplicable deficiencies, weak plants in spite of chemical fertilisers, and a lack of resistance to pests. An increase in infestation is a consequence of this biological imbalance. Farmers should consult their colleagues who have realised the necessity of biological cultivation. Divorced from thoughtless practises, from ignorance and from systems without a future, biological cultivation now knows where it is going.

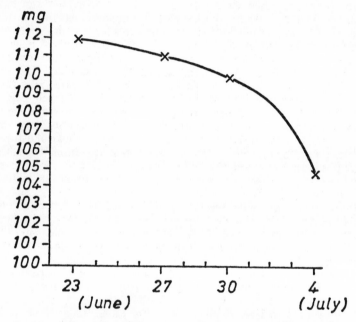

Fig. 20. Phosphorus rate variation in lentil germination (Colin thesis). There is a decrease of more than 5 per cent in total phosphorus within twelve days. (P value for 400 grains.)

The mechanism of these biological transmutations allows an understanding of what the soil needs—provided that it is a " live " soil, rich in microorganisms whose proliferation is possible. If the soil is already too much altered by chemical abuse, it must be biologically reconstituted by methods well known to the promoters of agricultural associations for biological cultivation. Soil reconstitution takes a long time, especially in the absence of humus, the essential part of a living soil.

Classical soil scientists and agronomists attached to the dogma that biology equals chemistry, cannot conceive that all that is in plants has not been put into the soil. They are not the people to advise farmers; they should be guided by the enlightened and intelligent agriculturists who have long recognised the division between a purely chemical and a biological agriculture. They might then achieve their own conversion, and carry out some of the experiments described in this book for themselves. If they are men of good faith, they will admit their past errors; but one doesn't ask that much—only that they act.

Farmers know that chemical fertilisers are seldom assimilated direct. Plants only utilise them through the activities of microorganisms, the moulds and microscopic algae, the bacteria and the actinomycetes (which live like bacteria but reproduce like moulds, and like them have a filamentous structure). In a fertile soil there are between hundreds of millions and thousands of millions of these organisms in a cubic centimetre. The activity of higher forms of life, annelids or earthworms, etc., must not be forgotten, and being more highly developed they are more vulnerable and the first to disappear as a soil dies. These macroorganisms are all aerobic; the majority of microorganisms are also aerobic, the proportion of anaerobia being much lower. Bacteria are mostly heterotrophic, that is they need organic matter to sustain life. Those living on mineral salts, the autotrophic, are rarer, but they have an important role.

Without the intervention of microorganisms normal cultivation is impossible. Therefore the top-soil must be aerated frequently, and the subsoil should be worked every five or six years. The soil must contain organic matter. A good mechanical soil treatment can multiply the microflora five or even ten times in one to three weeks, but this in itself is not enough; the most abundant microorganisms grow within narrow acid-base conditions (the pH), generally between pH 6.5 to 7.5, or at most between pH 6.0 and 8.0. The case has been mentioned[1] of manganobacteria, where activity decreases above pH 7.3 and stops at 7.8; in an acid soil it stops at 6.3.

So even if, chemically, a soil contains manganese, or if this element is given to it, it has no effect if the soil has too much lime or is too acid. Chemical analysis will not indicate any deficiency, but the plant will clearly show it. This type of deficiency has been especially studied in relation to oats, pear trees etc. showing Mg deficiencies in certain soils etc, where the condition seemed inexplicable since chemical analysis revealed the presence of manganese, which proves that chemical soil analysis gives insufficient information.

It is a pity that agriculture has no specialised service to supply

[1] See *Low Energy Transmutations*, p. 242

strains of microorganisms commercially (the project exists but is in early infancy as yet). The cultured strains could be diluted in a nutrient medium and spread on the fields (or injected at the base of trees) where a vital soil level has to be re-established—after preparatory treatment to ensure the correct pH and sufficient organic matter. The same would apply to soils having the correct constituents which were, nevertheless, unassimilable in the absence of microorganisms to effect the necessary transmutations or chemical oxidation-reduction reactions. A scientific agricultural body conducting research in this subject would have to determine which chemical products are truly of value. The decision should not be left to the fertiliser manufacturers. Besides, the usefulness of a product depends on an analysis of the soil, since one must find out what is in the soil, in physical and chemical forms. Even so, what is missing in the soil should not of necessity be supplied; a particular chemical may not be assimilated by the plant, or only poorly so, because it is an end product, and incapable of being biologically transformed.

In this way one is led to compare inorganic and organic forms of nitrogen, but nothing can provide urea in the soil if the soil has insufficient specific bacteria for urea decomposition (two which must be present are: *Micrococcus urea* and *Urobacillus duclausi*). The form in which lime is supplied is also important, it being generally of low solubility. An inert mineral form of lime is only slowly transformed; its crystaline structure can change, or disappear altogether, giving a metamorphic cement. Such lime will have become compact or even finely pulverised, as its particles are very small, about 0.1 mm, or even 0.05 mm. But these are still very large in comparison with bacteria, and it would be better to have porous particles presenting a large surface area for action by bacterial enzymes. We have already noted the difference between mineral calcium carbonate and the calcium carbonate in fresh lithothamnus. In the latter, the cells are three to nine microns in size and are surrounded by a calcareous or, to be more precise, calcareous magnesia membrane in which the pores and communication canals have a diameter of about 0.1 micron (1/10,000 mm).

A large part of agronomic research must therefore be reconsidered, and it is to be hoped that pressure by associations for biological cultivation will be such that revision of earlier concepts will be accepted in official quarters.

Already, bulletins, journals and reviews published by these associations contain articles on biological transmutations, and reports of new observations which confirm them. It is not possible to cite them all here[1] but one example appeared in *Nature et Progrès*, September 1965. E. Cussonneau reported the analysis of P_2O_5 in two apparently identical soils. One received a fermented compost, the other received

[1] Some appear in *The Journal of The Soil Association* (Great Britain); in France in *Agriculture et Vie*, edited by R. Lemaire and Sons; in Italy in the Bulletin of *I.B.I.N.*, edited by A. Ghiotti, etc.

farmyard manure; in other words the first received no phosphorus but the second did. The analyses were made monthly; here are the results:

	Jan.	Feb.	Mar.	Apr.	May	June	July	Aug.	Sept.	Oct.	Nov.	Dec.
Without phosphorus	14	35	40	30	20	27	23	20	15	30	27	32
With phosphorus	17	16	18	27	17	16	20	14	16	17	15	13

A total of 314 mg for the first, and 205 mg for the second.

He concluded: " Therefore the soil containing the greater amount of phosphorus was the one without any external supply of this mineral. A miracle of the living soil ". Dahiez, of the Soils Laboratory in Brittany, reports an increase of phosphorus in plants arising after giving sulphur to the soil; and M. Tavera reports that soil on a farm in Languedoc had its potassium content increased without any application of potash fertilisers, a soil which underneath was saline. Again in Holland, Professor Grasshuis, director of the Institute of Research into Animal Nutrition, and cited by Ch. Bauer in *La Préscience de Rudolph Steiner* reported that the abuse of calcium fertilisers resulted in 50% of soils having too high a pH; leading to a shortage of Mn in 30% of the fodder samples (less than 100 p.p.m. Mn in the dry matter). In 85.5% of the samples there was also a copper deficiency (less than 3 p.p.m.) which affected the blood serum of lactating cows, 57.3% of which had less than 0.30 mg Cu per litre. These deficiencies led to sterility in the animals; on 30% of the farms less than 60% of the cows were fertilised at the first insemination, and in 20% of these cases the fertilisation rate was under 50%. The copper and manganese deficiencies (i.e. microorganism activity linked to iron according to our studies) lead the author to conclude " excess calcium causes a deficiency in copper and manganese ".

The reader of this book will now understand the causes for these observations with reference to classical biological chemistry, and one has a better appreciation of the observations made by Pochon and reported in his *Traité de Microbiologie des Sols*. In a study of excretion of $CaCO_3$ by certain glands of the earthworm he compares the findings of various authors who made trials in identical soils, with and without earthworms. After the trials, soils with worms contained more calcium, magnesium and phosphorus than those without worms. (As we know the bonds between the three elements, we can say that the earthworm's need of silica and of clay-silica is further evidence of silica changing to lime through the presence of annelids.) One can also understand the astonishment of classical agronomists who found that nitrogen supplied to the soil disappeared at a rate of 15 to 20 tons per hectare and sometimes even more; it was impossible to trace this nitrogen in spite of taking all precautions. They had not thought that $N_2 := C + O$.

89

One is amazed at the inability of most agronomists to foresee the results of their ill-considered action when resorting to intensive use of chemical fertilisers, particularly its exclusive use. Mineral products (potassium, phosphorus, etc.) can enter only when they form part of the composition of organic molecules; this involves biological activity, which begins in the soil and is carried out by microorganisms.

To give plants chemical fertilisers only, is simply giving them drugs to achieve higher crop yields—for a time. It is just as if we had had our appetites stimulated by aperitifs, but not followed up by a meal. The necessary microorganisms must be allowed to live. In order to live they must be able to breathe, and give off carbon dioxide gas, CO_2, which is not their source of carbon but a noxious waste product. They must get their carbon and their oxygen from the soil.

Autotrophic bacteria can obtain their carbon and their oxygen from mineral molecules (carbonates) but these bacteria, thiobacilli etc., are not plentiful; the great majority of microorganisms (hundreds of millions per gram of soil) can only use carbon and oxygen from organic matter.

Therefore as long as the microorganisms have organic substances available, their growth and reproduction are accelerated and this results in improved crop yield. But if only chemical fertilisers are applied to the soil, without any accompanying organic material, in a few years the soil becomes exhausted and is without life. The U.S.A. have experienced this on a large scale, but, in more traditional France the position is not quite the same because, fortunately, we hesitated to break with the past. Along with increased quantities of chemical fertilisers, a certain amount of organic manure is still being given, but there are places where this practice has been forgotten.

The evidence in favour of supplying organic material has been overlooked by agronomical theorists (by too many of them). They have ignored or forgotten that carbon is essential to living organisms, whether soil bacteria, moulds, actinomycetes, annelids, nematodes, amoeba, protozoa etc., in all macrofauna and microfauna of the soil; macro- and micro-fauna are indispensable for assimilation of food materials through the roots. Without doubt, chemical fertiliser manufacturers will attempt to prepare organic materials synthetically, but will such preparations be economical to use and will they be biologically valid, because it is not enough merely to have worked out the chemical formula. (Molecules are dextro- or leviro-.) At all events, it now rests with the farmers to halt the slide towards catastrophe by ensuring that their allies, the microorganisms, are nourished by carbon and organic matter and should no longer be drugged, because someday they will become exhausted and die; their death will mean sterility of the soil they lived in.

However, we must guard against over-simplification and rigid doctrinal points of view; life is too complicated for us to be tied to a few elementary principles. To close this chapter I give an example;

it is not directly related to agriculture but completes the reference to small forms of animal life in the soil. We have seen that earthworms are able to transform the silicon of aluminium silicate (clay) into lime (calcium carbonate).

A systematic investigation of cave microfauna was carried out by the biospeleology laboratory of the *Centre National de la Recherche Scientifique*, installed in the Moulis grotto in the county of Ariège. Thus the study was made in the environment where the fauna had developed. Some of the results were published in the March 1966 number of *Science et Avenir*, and the following is an extract: " Though specimens of young Niphargus (a species of shrimp 15 mm long living in the clay of the caves) survive without food for several months without development before dying, some kept in tanks containing clay grow normally in the absence of any other external food supply. Furthermore the same shrimps deprived of clay and fed exclusively on fragments of meat still grow but never survive longer than a year.

Therefore clay seems indispensable for growth of these young creatures, and may constitute an exclusive diet for certain species of worms and molluscs. The secret of their survival in a medium totally deprived of any source of vitamins is that the clay is vitalised by the bacterial populations which it contains . . . the vitamins present in the soil are produced by certain microorganisms . . . in the absence of light certain bacteria synthesise the principal vitamins by chemosynthesis . . . autotrophic bacteria in a subterranean environment are not pigmented, but they carry out simultaneously an intensive vitamin production and a fully developed endogenous cycle enabling them to lead a wholly autonomous existence."

The conclusion was therefore reached that " life would not be possible in the caves without clay."

The investigators showed that, in the example given, wet clay (in caves the humidity is almost 100%) is necessary to life and can adequately support it.

It was not established that the shrimps can live in clay by themselves; apparently the many bacteria also present are the agents that prepare the food for the shrimps. However, this only shifts the problem, since at whatever stage the change occurs (in the shrimps, or by the bacteria) it is evident that clay is the primary material

These bacteria are autotrophic, subsisting without organic matter. They are also anaerobic, that is they do not need contact with air. So they cannot obtain their carbon from the carbon dioxide in the air, nor from the clay which has neither mineral nor organic carbon. If an organic food material is added, the shrimps weaken and die.

But the shrimps breathe, the bacteria also. Respiration involves emission of carbon dioxide, CO_2, which is a waste product. Oxygen can come from water, but where does the carbon come from? No matter what way the problem is considered, it cannot be solved by classical chemistry. The sole explanation is that the autotrophic

91

bacteria (silicobacteria) break down the silicon of the clay into C + O.

Only biological transmutations furnish a valid explanation for the origin of carbon in a substance which does not contain it in a chemical form. It is certain there are other biological transmutations enabling shrimps and other small forms of animal life to find in the clay (direct or through microorganisms) minerals indispensable to them, especially sodium, magnesium, phosphorus, sulphur, potassium and calcium. Within this closed system these minerals are not renewed, nor are they, in general, present in clay at all.

Finally, before going to the next chapter, I should like to pass on some information from a teacher of agriculture. He drew my attention to a report to the Académie d'Agriculture, January 12, 1966, giving the results from a study of potassium balance in agriculture, made at the East Central Agricultural Station C.R.A. at Dijon. This study shows that the difference made between "exchangeable" K and "fixed" K is not justified; but the "export" of fixed K by plants is sometimes twice as much as that of exchangeable K.

We know that exchangeable K is usually the K soluble in oxalic acid, but this does not mean that it is usable K in a form which the plant can absorb, because bacteria are able to utilise the potassium from molecules which are insoluble in oxalic acid (which do not produce potassium salts of oxalic acid). This is why, in my books. I give total K, to avoid the so-called "refutation" of those who maintain that there is a transfer of "fixed" to "exchangeable" K, and that an increase of K in plants—beyond that from exchangeable K in the soil—is balanced by a decrease of acid-insoluble K due to undetermined external causes. (They have not made the relevant chemical analyses . . .) The report of January 12, 1966 to the Académie d'Agriculture is eloquent on this subject.

In two separate plots of soil were determined:

(a) K supplied in the form of fertiliser
    + decrease in soil content of K exchangeable against NH_4
    + decrease in soil content of fixed K (dosed by tetraphenylboron)

(b) Total K in the harvested plants

The total figures in kg/ha were:

(1) In the first plot, (a) was 1534.5; (b) was 1695, representing a surplus of 10.45%

(2) In the second plot (a) was 1629.5; (b) was 1772.5, representing a surplus of 8.77%

The average excess K taken away by the plants was 9.66, or to the nearest whole number, 10%.

There was no attempt to explain these results, strange though they were from the point of view of classical agrochemistry. The plants had taken away more potassium than was supplied to the soil, taking into account the reduction in total soil potassium, assimilable or not, present before any addition of potash fertiliser.

Where did the potassium come from? The reader knows the answer. This report coming from our leading agricultural body is one more confirmation of what was said at the beginning of the present chapter, part 3 Agriculture-Agronomy, and further substantiates that K is produced by plants, that is, by biological transmutation.

Chapter 9

GLIMPSES INTO THE FUTURE

> "It is a bad principle to deny a fact simply
> because we do not understand it."
>
> PROF. G. PICCARDI.

It is not possible in this short book, intended as a simple introduction, to review all the inferences of biological transmutations. We have outlined some of them in our other books; here is a partial resume:

1 CREATION OF THE EARTH (AND PERHAPS THE OTHER PLANETS)
SILICON—ALUMINIUM

It is generally believed that planets very distant from the sun (Jupiter, Saturn, Uranus) have an ammoniacal atmosphere, or rather one composed of nitrogen. Is this because of reduced solar activity with such great distances (ultra-violet rays, proton showers, etc.)? Do non-biological transmutations take place there very slowly, and are these distant planets in a less evolved state than our earth?

We can assume that more than 4 billion years ago our planet was still a gaseous mass rich in nitrogen. The nitrogen, under the effect of solar rays, cosmic particles etc., would "condense" to silicon according to the reaction already given:

$$2 \text{ x } \overset{14}{\underset{7}{\text{N}}} \; : = : \; \overset{28}{\underset{14}{\text{Si}}},$$

since nitrogen is normally in the form of a molecule containing two atoms of N.

One could envisage similar developments (see *Low Energy Transmutations*): the microscopic algae, *Diatomaceae*, surround themselves with a siliceous membrane, and form the silica in water where it is virtually absent (they can fill lakes, obstruct rivers, etc). It is naive to say that this silica is "fixed" by algae from the silicon in the water, since the algae eventually exceed in volume the water in which they live. Besides, silica is only slightly soluble in water, but that does not prevent large deposits of it being formed (also called diatomaceous earth, kieselghur, infusorial earth, etc.).

94

Silicon can lose a proton:

$$silicon - hydrogen = aluminium.$$

In this way silica and alumina were formed on the earth's crust, together with aluminium silicates, clays and all the elements which constitute primary earths.

The bond between silicon and aluminium is not a flight of fancy. In *Low Energy Transmutations* there is a report of experiments on rabbits: after adding powdered aluminium to the rabbits' food, more silicon was found in the ash of their incinerated carcasses than in that of control rabbits fed without added aluminium.

And we have given an explanation of how the earth passed from the Primary to the Secondary era.

A further point comes to mind: we might well ask, in view of the number of occasions where silicon = carbon + oxygen, if this accounts for the properties of the silicones. Organic silicon compounds have been obtained which show analogies with organic carbon compounds. Is this because silicon is basically only a " carbon complex "? Moreover, silicones are not formed in a medium very rich in oxygen: is this because oxygen is already bonded to the carbon in silicon?

2 COAL—MINERAL OIL

I will only mention here that the property of matter, as shown in reactions I have proposed, forms the basis of new hypotheses of coal and oil formation, which could in turn lead to new methods of prospecting (I have briefly described these hypotheses in *Natural Transmutations*). In this present work I have outlined in Chapter 4 the process of the formation of carbon of mineral origin, and shown that certain rocks, such as ampelite or clays used as "foundry black", can have up to 20–30% graphite. Graphite is found in rocks which antedate the appearance of vegetation on earth, so carbon could well have an origin other than organic.

3 SULPHUR

Sulphur is bonded to phosphorus (+hydrogen), but it can be seen that my proposed structure of the addition of phosphorus and hydrogen, leads to the same structure as a double oxygen atom:

$$2 \times {}^{16}_{8}O \quad := \quad {}^{32}_{16}S.$$

Therefore the origin of sulphur, which has always been a mystery, could well be a " condensate " of oxygen. Chemists, of course, know what these two elements have in common. Volcanic sulphur, and that in the natural gas of Lacq etc. could not have originated in any other way, neither could the sulphur which gave rise to the thiobacilli. (Also, the loss of a proton from the double oxygen nucleus would give phosphorus, an element which can also come from magnesium with lithium.) In nuclear acids, between two chains of proteins, there are phosphorus or sulphur bridges, and biological transmutation explains how Nature can cross from one to the other.

4 OXYGEN

Oxygen would not have existed when the earth was being formed, and I propose the hypothesis that it came from nitrogen according to the reaction already given: $2N := : C + O$. The pre-existing nitrogen at the time of the first solid condensation of the earth's crust would have given silicon or carbon and oxygen according to radiation intensity or perhaps radiation effects. As the earth grows older, does its atmosphere become richer in oxygen? Does life become more active? Is evolution more rapid? Consequently, on planets which, on account of their proximity to the sun and strong irradiation, have aged more quickly there is no more nitrogen N_2, or very little, it having almost disappeared by decomposition into $C + O$, leaving an atmosphere rich in CO_2. Is not this the case with Venus? The atmosphere of Mars can also be explained in the same way.

5 MAGNESIUM

Carbon can give magnesium:

$$2 \times \quad \begin{matrix} 12 \\ C \\ 6 \end{matrix} \quad : = : \quad \begin{matrix} 24 \\ Mg. \\ 12 \end{matrix}$$

This is not a matter of imagination. Magnesium has been obtained in a particle accelerator by bombarding carbon with a carbon nucleus.

Might this not explain what geologists term SIMA (combination of the first two letters of SIlicon and MAgnesium) which, when under the earth's crust, was "SIAL" (SIlicon and ALuminium)? The physico-chemical composition of SIMA is hypothetical (very deep drilling is planned in order to study this transitional geological layer).

In the earlier state of the earth's development the reactions were: $2N := : C + O$ and $2N := : Si$, which would explain the presence of more magnesium in the Precambrian ($2C := : Mg$), that is in the rock preceding the Primary era. Two carbon nuclei united to give magnesium, while the oxygen available for the chemical reactions united with magnesium and silicon. The nitrogen became either $C + O$ or Si, depending on the incidence of radiation showers, which varied with the earth's rotation.

6 PHILOSOPHY—METAPHYSICS

Everything concerning the origin and evolution of our planet requires reappraisal; new horizons will be opened up, not only for geologists but for philosophers and metaphysicists. The latter will also have to consider that, contrary to accepted statements, the phenomenon of life is not exclusively a matter of chemistry. There is chemistry in life processes, and obviously I would in no way deny the importance of chemistry and biochemistry in all living organisms, but life is not just physics and chemistry; there are other phenomena, namely biological transmutations.

All this does not mean that the above is one step forward in understanding what life is. On the contrary I would say that it is one more step which reveals that life is still more complex than biochemists think. There is now definite proof that something apart from chemistry is involved. Let us beware of generalisations and simplifications, and let us remain modest, acknowledging our ignorance.

Claude Bernard has said: " When we propose a general theory, the only thing we are sure of is that all theories are false . . . for with the growth of scientific knowledge they must be modified, the more so in the use of less highly evolved sciences. What characterises a real scientist is not making discoveries, in which chance can play a large part, but discovering the laws of phenomena."

Chapter 10

CONCLUSION

> "That a novice should dare to contradict the
> Masters, that he should even more foolishly
> try to convince them, reveals a remarkable
> misunderstanding of the world and a juvenile
> obstinacy."
>
> GOETHE.

The quotation from Goethe shows it is always difficult to make the "Masters" accept what they did not find themselves. Pusillanimity of spirit exists in many men, even among scientists. Knowledge itself does not change the spirit, neither does it bring intelligence. The two are quite distinct, I do not see that our contemporaries are any more intelligent than were the people of Athens or ancient Rome.

Excessive specialisation leads too many scientists to see nothing outside their own field. Today, too few keep their critical faculty intact or avoid a blind adherance to dogma. It is clear that science today is a trade; only a minority of those who live by science have the true scientific mind in which facts alone matter.

In the preceding pages I have presented some facts that no one has been able to refute or contradict. I have mentioned some of the great names in science, people who stand far above the anonymous crowd of specialists mass produced by our educational system to meet our present technico-economic needs. These prominent scientists have given me support and encouragement and have helped me to publicise my work by discussion, by their own work, and by their lectures and published articles. For these scientists only facts are of value; and since the facts can be demonstrated and proved, they can only be disputed by people who prefer to appear unconvinced.

One can argue over mechanisms, since we do not know how to explain all that takes place inside atomic nuclei: everything said about the forces involved and energies set in motion is so far only hypothetical. But hypotheses are applied to many phenomena. Even those who state certain scientific principles often ignore the original hypotheses and simple statements and deceive themselves by words— words often given as a valid explanation. There are many phenomena which, to satisfy our Cartesian spirit, we can "explain" only by words. But such a practice reveals our profound ignorance of the causes of many observed facts, especially in biology. In this science there are numerous interacting parameters. We are no longer dealing with sciences termed "exact" just because they are less complicated.

Thus medicine is often termed an "art" because those active in the simple sciences cannot imagine any science other than their own (biology is too complex and is beyond them) . . . Everything relating to life is very finely balanced; for example, agriculture is the application of centuries' old observations, and has a tradition that should not be presumptuously discarded in the name of so-called "modern science".

To be sure, in some fields modern science has made undeniably rapid progress, but let us not be so fatuous as to believe it has the answer to everything, and that by following certain theories (often temporary) which disregard established practices we can foresee what will take place.

I hope that the reader who has read and understood this book will be convinced of the existence of the phenomenon of biological transmutation which cannot be denied, and that therein is the key to a host of relating explanations for the many observed facts, which has been beyond the reach of classical science.

One should bear in mind after reading this present work that the phenomena of transmutations of elements are closely tied to the phenomenon of life (which does not rule out other possibilities as a result of powerful physical forces). However, one must not suppose that the reactions given in the preceding chapters take place automatically and are always and everywhere valid. They occur only under special circumstances, induced by enzymes secreted or synthesised only under specific conditions, such as an abnormal rise in temperature.

The task of the doctor, biologist, veterinarian, agronomist or stock breeder is therefore much more complicated than has been admitted. For the chemical "treatment" of humans, animals and plants there must be present in the cell or in a microorganism an enzyme catalyst.

But the simultaneous existence of a transmutable element and appropriate microorganisms or enzymes is not always sufficient. For example, in soil microorganisms must have physico-chemical conditions favourable for their proliferation. In *Low Energy Transmutations* I have shown that the microorganisms which assure the manganese-iron bond have very narrow optimal limits of temperature and, especially, of soil pH (= acido-basicity).

An abundance of bacteria in a dormant or enfeebled condition in a soil too acid or too alkaline will not be effective, and there will be a deficiency of manganese in the plant even if there is a large amount of it in the soil.

Each case must be studied individually, extrapolation will not work, because in biology nothing is simple; it is too complicated for those who are addicted to simple matters such as "exact" sciences.

Chapter **11**

AN EXPOSITION OF ATOMIC NUCLEI

> " There are heretical facts. Such facts, be-
> cause they put questions and because they
> disturb the intellectual comfort of many
> people, must be more fully explained than
> others."
>
> F. BRUNO.

WHAT IS ACCEPTED

The internal mechanism of transmutation of elements by biological action is unknown; likewise the mechanism of transmutation of a radioactive element. We know the result but we do not know what takes place inside the atomic nuclei. For example, no one knows why a radioactive potassium nucleus ($^{40}_{19}$ K) " explodes " at an indeterminate time, leaving behind argon which has 18 protons instead of the 19 of potassium, one proton of this last becoming a neutron. The natural radioactivity can be effected even at very low temperatures.

By a method well known in nuclear physics, an artificial transmutation can be produced by shock of a slow neutron (thermal neutron). This also occurs under very low energy; an isolated thermic reaction at 300°K or 27°C has an energy of only about 0.038 eV. This neutron is able to effect fission of uranium-235 or plutonium-239. Compare the chemical reaction $C + O \rightarrow CO_2 + 94,000$ calories or 4.1 eV. In this case the energy is for one mole; in the previous one it is for a single neutron.

With elements of lighter atomic weights the reaction follows the formula:

$$^{A}_{Z}X + ^{1}_{0}n \rightarrow ^{A}_{Z-1}Y + ^{1}_{1}H + Q$$

and in several types of reaction protons are expelled. For instance, we can go from sulphur-32 (stable) to phosphorus-30 (radioactive) by means of a fast neutron; one proton is expelled since we go from $_{32}S$ to $_{31}P$.

100

If the ' projectile ' is an $\propto$-particle there is the same expulsion of a proton. Example:

$$\begin{matrix} 27 \\ Al \\ 13 \end{matrix} + \begin{matrix} 4 \\ He \\ 2 \end{matrix} \rightarrow \begin{matrix} 30 \\ Si \\ 14 \end{matrix} + \begin{matrix} 1 \\ H \\ 1 \end{matrix} + Q$$

and again the reaction:

$$\begin{matrix} 14 \\ N \\ 7 \end{matrix} + \begin{matrix} 4 \\ He \\ 2 \end{matrix} \rightarrow \begin{matrix} 17 \\ O \\ 8 \end{matrix} + \begin{matrix} 1 \\ H \\ 1 \end{matrix} + Q$$

these representing the first experiment on artificial transmutation which was carried out by Rutherford.

Many other examples could be given indicating presence in the nucleus of less firmly bound protons capable of expulsion under certain shock. Many physicists accept the presence in nuclei of weakly bound protons which can leave or enter the nucleus. This cannot be directly proved; at present we have no direct method of investigating the nuclear structure. The internal structure of nuclei can only be subject for hypotheses, based on indirect results and deductions from observed effects. Here Bayes' law applies; this states that starting from causes we are able to predict, under identical conditions the effects cannot be reversed. However, our knowledge of the atomic nucleus is still only hypothetical. Many of the hypotheses are invalid after lasting only a few years.

But we can accept the presence within the nucleus of some protons which have a partial freedom of movement. For such protons all hypotheses of average binding energy of the nucleon are invalid. If a calculation is made starting with the expulsion of these protons the value cannot be extrapolated to other nucleons. One can therefore state that this concept, which is regularly given in all nuclear physics textbooks, has no meaning. The Nobel Prize winner Pauling, in 1964, showed that if a uranium-235 nucleus splits in two parts ($\frac{3}{5}$ of nucleons in the heavier part, and $\frac{2}{5}$ in the lighter) it is because the nucleus is not an assembly of nucleons bound to each other by an identical energy (otherwise fission would follow the law of chance); the nucleus is heterogeneous and the nucleons are energetically bonded to each other in separate groups, but these groups are more loosely bonded between themselves. We have called such groups " nuclearons " = nucleon groups.

We published this concept of nuclear heterogeneity in 1960.

OUR DIAGRAMS AND THE THREE KINDS OF BIOLOGICAL TRANSMUTATIONS

A large amount of converging data took us to the belief that

101

nucleons first group themselves into a "core" with a maximum of 5 protons (we are not considering neutrons, as in the various isotopes, but only the protons which characterise the nature and the name of an element).

Protons exceeding 5 (therefore beyond boron, $_5$B) in number are arranged in a second orbit outside the core. Those in this second orbit seem to be less firmly bound than those in the first and do they perhaps constitute the mobile protons? In a nucleus made up of several of these primary cores these protons would be able to move from one core to another. Such cores would therefore be a type of "sub-nucleus"; the nucleus itself would not be homogeneous but granulated, each "grain" being a nuclearon.

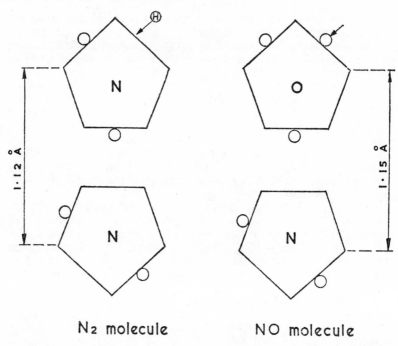

N₂ molecule   NO molecule

Fig. 21. Left, the two nuclei of the N$_2$ molecule. One (H+) proton attaches itself to the upper nucleus of this molecule. We then have the new molecule on the right, which is NO.

Thus carbon $_6$C would have a core of boron with one external proton; nitrogen $_7$N with 2 external protons; oxygen $_8$O with three external protons, etc. up to a maximum of 5 in the external nuclear layer. Beyond Z = 10 (10 protons) there would be two groups (two nuclearons) bonded together but remaining distinct from each other. Thus sodium $_{11}$Na (5 + 6 protons or $_5$B + $_6$C), followed by mag-

102

nesium, $_{12}Mg$ (6 + 6 protons or $_6C + _6C$) etc. Beyond $Z = 18$ there would be three "primary" groups or 3 nuclearons, for example $_{19}K$ and $_{20}Ca$).

We limit our examples to these reactions and exclude elements heavier than calcium which are less frequent in biology. We can now distinguish the three types of transmutations:

(1) BY DISPLACEMENT OF A PROTON INSIDE A MOLECULE

Take the molecule N_2. It is formed of two atoms of nitrogen (monatomic nitrogen is unstable, it is either bonded to one or more other atoms or to itself). In the molecule the distance between the two nuclei is 1.12 Å. (The third decimal place is doubtful, so we give only two.) Each atom has its own nucleus, and each nucleus is surrounded by its layer (K-orbit) of two electrons kept equidistant from the centre of the molecule by electrical attraction between positively charged nuclei and negatively charged electrons. This is shown diagramatically in Fig. 23.

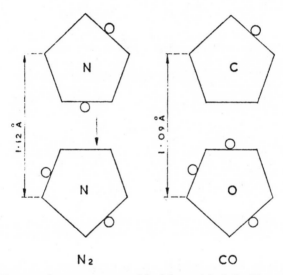

Fig. 22. Diagram of two separate nuclei: left, nuclei of the N_2 molecule; one proton of the upper nucleus travels in the direction of the arrow, and is captured by the lower nucleus. We then have the molecule on the right, CO.

The plane of rotation of electrons in the K-orbit is perpendicular to the axis joining the two nuclei. Observe the nuclei revolving one around the other so that relatively the electrons of the external layer (in the outside orbit) form an envelope which apparently revolves around the nuclei. (This could be compared to a circle of fire made by a person spinning round while holding out a flaming torch.) Of course this is only an illustration to help explain the phenomenon. The electrons in the K-orbit revolve with the nuclei inside the orbit formed by external electrons. A complete schematic representation would be similar to Figure 6.

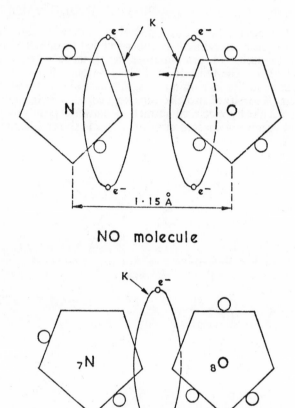

NO molecule

₁₅P nucleus

Fig. 23. Above, the NO molecule; the two nuclei approach closely, and we then have the P nucleus (below).

Let us assume that by an action involving energy (due to an enzyme) a proton becomes displaced (see Fig. 22 arrow) by vibrations reaching a resonance peak (vibration frequencies of nuclei inside a molecule are in the infra-red range). At the instant when the proton comes closest to the lower nuclear core, due to the effect of the electron layer K it is trapped by the nucleus. Thereupon the higher nuclear core has only one external proton while the lower core has three. The upper one becomes carbon, $_6C$ ($_7N - _1H := _6C$) and the lower one becomes oxygen, $_8O$ ($_7N + _1H := _8O$). The distance between the nuclei of C and O is then stabilised at an average of about 1.09 Å.

It is seen that no change has occurred in electron arrangement; everything has taken place within the electron envelope and inside the molecular orbit. There has been an internal movement of only one proton; the total number of protons, neutrons and electrons remains unchanged. There was neither an emission nor an absorption of particles. We have indeed a transmutation, as the nitrogen has become carbon and oxygen. The shorter distance between the nuclei indicates that the whole structure is now more compact and has a higher energy; in fact it had been necessary to supply energy. Here we are in a situation where a molecule of nitrogen is substituted by a molecule of carbon monoxide. This would explain the findings concerning respiration of nitrogen activated by an incandescent metal which leads to formation of carbon monoxide in the body.

It is evident that this phenomenon is not in accordance with the teachings of classical nuclear physics; but this does not imply that the problem of theoretical study of the energy involved is resolved. Physicists have advanced certain explanations—incomplete—which will not be considered here. They admit to being confronted by a new phenomenon.

(2) ABSORPTION (OR REJECTION) OF A PROTON

Let us consider again the diagram for nitrogen, N_2.

We know that in chemistry there are receivers and donors of protons. Therefore by the action of enzymes there should be the possibility of taking away or adding a proton. Let us assume a proton (coming from outside the molecule) manages to join the nucleus on the right; it then has a total of 8 protons. Therefore the nitrogen has become oxygen and the N_2 molecule has become NO; the distance between the nuclei, formerly 1.12 Å is now 1.15 Å (Fig. 21). (The energy is reduced and is exothermic.).

(3) APPROACH (OR SEPARATION) OF NUCLEI

(a) Let us look at the reaction $_7N + _8O := _{15}P$ (Fig. 23). We start with the molecule NO. By resonance due to external energy, when a certain high energy level is reached the nuclei (with their

K-layers) come so close together that the electrons of the K-layers repel each other. When one electron comes quite close to another revolving in an adjacent orbit (we have said that the K-orbits revolve between the nuclei) the two electrons form, in fact, two parallel currents, in the same direction, and attract each other by an electrodynamic interaction until the electrostatic repulsion becomes more powerful and breaks up the K-orbit. Pauli's principle is applicable here: the two electrons which repel each other are those of different spin and are thrown into a higher orbit. The total number of electrons does not change but there is a re-arrangement in the orbits other than K where two electrons of opposite spin remain; the re-arrangement in the orbits further away requires only a low energy.

But the two nuclei which came close together (sintering) are now fixed to each other by the nuclear force. The electrostatic force of the two electrons—which form the sole orbit K between the two " subnuclei "—also act on them. So there is a new arrangement formed of one nucleus of N and one of O practically united; this is the phosphorus nucleus ($_7N + {}_8O : = : {}_{15}P$). The reaction liberates an energy of several keV (according to a theoretical study by the physicist L. Romani).

We explain thus how a proton attached to a nitrogen nucleus becomes an oxygen nucleus: it means that if the second nucleus of the N_2 molecule gets another proton, there are in fact two oxygen nuclei: $_8O + {}_8O : = : {}_{16}S$ but the same scheme can also be: $_{15}P + {}_1H : = : {}_{16}S$.

(b) Let us now take the molecule of MgO.

We will thus extend the plan to a heavier element, one having a group of three attached nuclearons. We know that Mg $:=:$ 2 x C signifies that two attached carbon nuclei are equivalent to one magnesium nucleus (this has been confirmed in a particle accelerator by projecting a C nucleus onto a C target). In the molecule MgO, the distance between the Mg and the O nuclei is 1.75 Å.

By the action of enzymes there can be a nuclear attachment similar to that in the preceding case, with the O nucleus joining up with the Mg nucleus formed out of the two C nuclei. The result is calcium: $_{12}Mg + {}_8O : = : {}_{20}Ca$.

REMARKS ON THE CALCIUM NUCLEUS

We see that the calcium nucleus can be considered as formed by the union of one magnesium and one calcium nucleus, and we have described a number of experiments showing the origin of calcium starting from magnesium.

But of this group of three adjacent nuclearons, the two on the right, carbon and oxygen, may stay together; in which case $_6C + {}_8O : = :$ $_{14}Si$. In this grouping we can arbitrarily say that Ca is either Si + C or Mg + O. Experience tells us that nature is able to make Ca from Si; similarly it is seen that Ca is K + H, or K is Ca − H. This shows us that in fact the Ca nucleus is always the same; our system of artificial classification compels us to write different formulas in order to conform with scientific practices.

(4) EQUIVALENT MEANS OF TRANSMUTATION

Since protons possess a certain freedom of movement within the nuclei, at a given point in their rotation we can quite easily attach them to one or other " core ". The number of protons does not alter and there is no change in the total internal energy. What is shown in the diagram is an exact moment in the varying position of the protons. For example, we can likewise represent silicon by two N nuclei, or with a slight internal displacement of one of the protons during rotation, by C + O. Does this perhaps explain the presence of graphite in siliceous rocks, due to the Si nucleus splitting into C + O?

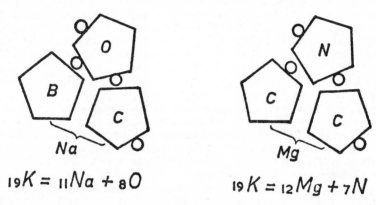

$$_{19}K = _{11}Na + _8O \qquad _{19}K = _{12}Mg + _7N$$

Fig. 24. Equivalent schemes: left, one of the O protons goes to B; O then becomes N, while B becomes C. We then have the equivalent Mg + N in the place of Na + O. (Note that if N picks up an extra proton from outside, it becomes O, and we would have $_{19}K + _1H :=: _{20}Ca$, which is also $_{12}Mg + _8O$, or again we have in total the protons of the NaOH molecule. We see that K is also $_{13}Al + _6C$.)

We say that if one proton is taken away, aluminium remains. Therefore starting solely from the N_2 molecule we have the elements to provide silica, alumina, aluminium solicates, clay, etc., all of which existed in Primary rocks. At a further stage were the Secondary rocks, since $_{14}Si + _6C :=: _{20}Ca$. We also mention the passage of iron to manganese by a simple loss of one proton. Experience shows this can be effected by the streptomycetes which can take away a proton from the iron. The reverse process occurs in seed germination, where $_{25}Mn + _1H :=: _{26}Fe$.

Let us take the example of an element with three nuclearons, potassium (Fig. 24). This could be $_{11}Na + _8O$ (and $_{11}Na :=: _5B +$

107

$_6$C); if one of the protons goes to B, then O becomes N and B becomes C. K is then represented by N + C + C. In fact both these tracks are possible in biology. It might also be said that $_7$N + $_6$C : = : $_{13}$Al which leads to $_{19}$K : = : $_{13}$Al + $_6$C, but this reaction has not yet been observed in biology. Theoretically it cannot be excluded.

(5) LAW OF EXCLUSIVITY

In the previous reaction, we have a case of exclusivity, a subject we have so far not considered. It was dealt with in our first book in 1962. For simplification we take only the protons. Neutrons should not be omitted; they have their own laws. For an unknown reason (probably because there is no equilibrium between spins in opposite directions when there is an odd number of protons) in order that the nucleus remains stable the number of neutrons must be higher than the number of protons. (If Z is an even number, stability is maintained up to Z = 20, which is calcium. Beyond this a higher number of neutrons is needed.) Only lithium 6 and nitrogen 14 are stable with equal numbers and odd numbers of protons and neutrons. They are called odd-odd, because both protons and neutrons have odd numbers. Higher than Z = 7, two odd numbers are no longer stable. The stable elements where Z is an odd number will always have at least one supplementary neutron. Commencing with chlorine where Z = 17, there could be either 1 or 3 supplementary neutrons making a total of 18 or 20. The total can never be an odd number, since this would mean an odd-odd (radioactive). Thus chlorines 35 and 37 are stable but chlorine 36 is not. As the nuclei become heavier, the number of supplementary neutrons increases.

Nitrogen 15 is stable (it is not an odd-odd); aluminium with an odd Z is only stable with 14 neutrons. Up to Z = 17, where Z is an odd number there can be no more than one supplementary neutron. In elements where Z is an even number there can be only one supplementary neutron for C, so we have the stable isotopes Carbon 12 and 13. (Oxygen can have two supplementary neutrons, giving stable isotopes 16, 17 and 18.) If aluminium is formed by C + N we have the following combination of nucleons:

26 (12 + 14); 27 (12 + 15 or 13 + 14) and 28 (13 + 15).

26 and 28 must be discarded as there are 13 protons and an even number of nucleons would not be possible without an odd number of neutrons. Therefore for aluminium the only stable combination is $^{27}_{13}$Al. (Similarly with Si, the only possibilities for obtaining

$$\begin{array}{cccccc} 27 & 28 & 1 & 29 & 2 \\ \text{Al are} & \text{Si} & - & \text{H or} & \text{Si} & - & \text{H)}. \\ 13 & 14 & 1 & 14 & 1 \end{array}$$

We now look at K, which might come from Al +C. As there are

108

both ^{12}C and ^{13}C we can have: ^{27}Al $+$ ^{12}C $:=:$ ^{39}K (possible and stable) or ^{27}Al $+$ ^{13}C $:=:$ ^{40}K (radioactive for a very long period: 1.3×10^9 years).

In the same way $_{14}$Si $:=:$ $_7$N $+$ $_7$N, and as we can have ^{14}N or ^{15}N there can be for Si either ^{14}N $+$ ^{14}N $:=:$ ^{28}Si, or ^{14}N $+$ ^{15}N $:=:$ ^{29}Si, or ^{15}N $+$ ^{15}N $:=:$ ^{30}Si; these are the only stable isotopes of Si. Similarly only ^{12}C and ^{13}C are stable, so Mg will be ^{12}C $+$ ^{12}C $:=:$ ^{24}Mg or ^{12}C $+$ ^{13}C $:=:$ ^{25}Mg or ^{13}C $+$ ^{13}C $:=:$ ^{26}Mg, which are the stable isotopes of Mg.

(6) AN EXAMPLE OF THE ENERGY INVOLVED

We have seen in the N_2 molecule that the distance between the two nuclei is 1.12 Å, and that it decreases to 1.09 Å in the carbon monoxide molecule CO. This implies energy to effect the change from N_2 to CO. We have also seen there is very little difference in ionising potential between these two molecules; the total energy bonding the electrons to the nucleus of 112 eV is higher in the CO molecule. Therefore there is only a slight difference in internal energy between these two molecules and in no way comparable with the tens or hundreds of MeV occurring in nuclear physics. In this reaction only one nucleon is displaced; the " lost " proton has moved from one nucleus to the other within the same molecule without disrupting the electron envelope. This transfer of a proton is apparently connected with a resonance effect since we know that the distance between the nuclei varies by an average vibration frequency of 3.5×10^{13}. It has been established that displacements of protons within a molecule are not instantaneous but are of the order of 10^{-13} seconds. As these two separate values are within the same range, could this vibration in fact be the " transfer wave " for the proton?

We say that if a transmutation takes place by incorporation of a supplementary proton, this proton can only come from outside the molecule, that is from another molecule. This means travelling a longer distance at a reduced speed because it has to pass the electron sheath (the molecular orbit); in this case it has been established that the reaction takes about 10^{-9} seconds, and lasts 10,000 times longer.

From the foregoing example it will be realised that an energy value of 112 eV is needed for a transmutation effected by internal displacement of a proton. In fact we do not know what happens in the nuclei; there could be a recurrent action caused by enzyme activities; or this displacement could be a result due to violent action between opposing forces, such as takes place in chemical reactions. In some cases we are beginning to assess the high values of energy produced by opposing forces in very short time periods, and recording them by ultra-fine measuring techniques (up to 10^{-6} seconds). But these high energy values are rare and the most frequent ones seem to be in the range of 10^{-9} seconds which is already evoked by the transmutation. What we write down as a chemical reaction is only a concept; it does not show what actually takes place. We observe an initial stage and a final stage and we know nothing of what really happens. We can only measure the total energy liberated by opposing forces and not the energy liberated at any particular moment.

In the transmutation reaction just given we stated that when the molecule N_2 changes to CO there is 112 eV energy more at the end stage than at the initial stage. This is all we can say, but we can also say that when one element increases, another decreases; by this we are led to think that there are probably two nuclear reactions of opposite directions shown only by the ensuing difference in energies of each of them.

By the above examples we have wished to show the following: that only little energy has been expended; that nuclear reactions of opposite directions are linked and nearly compensate themselves; we can assume that they occur in the manner of oscillating internal discharges caused by a recurrent very low energy which leads to a vibration resonance of the molecular nuclei. In our view, these reactions are quite distinct from those applied to isolated atoms by classical nuclear physics. An altogether different phenomenon is thus involved; as the acting energies are out of proportion thus confirming that there is a quite different process. The laws of this phenomenon should be specified on the basis of theoretical physics; several physicists are already working on this problem.

CONCLUSION OF THE PRESENT CHAPTER

This book presents only an imperfect view of the reasons why the physics of living matter is different from that of dead matter. But we must draw to a close. We do not claim to have formulated a theory of biological transmutations; but these transmutations exist; we only point out a possible line of theoretical research, and without claiming to have solved everything.

In this chapter our only aim was, in the face of our ignorance of the internal nuclear structure, to present certain hypothetical concepts which might bring about a better understanding of the mechanism involved in those structural changes.

We indicate examples, means to materialise what is inaccessible to our senses. Other examples could be quoted and this we have done in various works, but it is essential to note that the atomic nucleus is not a homogeneous collection of nucleons; we have written about it since 1960, as there is no other way of explaining the transmutations we have observed through irrefutable experiments: there are preferential collections strongly bonded which cannot be separated by enzymes; for instance: Oxygen forms a powerful block which passes from one nucleus to another resulting in a large number of transmutations with $\pm$ O (and with the " elementary " particle, the proton,

±H). We do not intent to give a complete study of these theoretical aspects with this book; but it is absolutely necessary to understand the principal reactions recorded in order to know when and how to apply them.

Chapter 11

THEORETICAL DISCUSSION

"What right have you to say to a fact: go
away?"

VICTOR HUGO

"A reasoning can never be opposed to a
fact."

LACORDAIRE

"One must be particularly obstinate to argue
against facts."

PROFESSOR DELBET

The reader may omit this final chapter. It adds nothing by way of
practical information to the subject of transmutations as already considered, but it should serve to prepare thinking persons to refute
opponents; though generally such opponents are victims of a system,
and it has been said that their opposition will only disappear when
they do. Also, as the wise La Fontaine has said:
"Est bien fou du cerveau qui prétend contenter tout le monde et
son père." ("Whoever claims to please everyone and his father is a
fool.")

On the other hand, he who objects without prejudice is entitled to
explanations, because his objections are usually based on *a priori*
judgements. He has not made a serious study of transmutations, but
has only heard of them in conversations or on the radio or through
incomplete and simplified written articles.

Professor A. Furon of the Sorbonne wrote in *Revue Général des
Sciences*, April 1963: "the rapid advance of scientific knowledge
precludes all orthodoxy and conformism."

We should be surprised at nothing, as long as what is reported can
be demonstrated to be so; it is only the results that matter, not the
theories.

The preceding pages have shown that transmutation of elements
in biology is in no way opposed to chemistry. Chemistry is a science
which deals with displacements of electrons in the peripheral atomic
layers; it is the science of molecules, not of atomic nuclei. The phenomenon I have demonstrated involves an alteration in structural ar-
rangements of atoms induced by enzyme activities in living matter. It
takes place within the atomic nuclei; therefore it is a new science,
quite distinct from chemistry. Chemistry is only a visible extension of

this new science to our perception and it is the application of chemistry, the final stage, that establishes that transmutation occurred.

The objections of chemists can therefore be easily met. They are mistaken; the phenomena studied—and the chemists should read this book, in fact they should be obliged to read it—will show them that chemistry is not in question, that transmutation involves a field of study foreign to them, and that I reject no aspect of true chemistry which is strictly limited.

On the other hand, there are opponents amongst atomic physicists. This young and presumptuous science also covers atomic nuclei; therefore there is a relation between its field and the one with which I am concerned. But the physicists are also mistaken; they postulate by calculations which they desire to be immediately accepted. Unfortunately, none of their calculations are valid, because when applied to biological transmutations their underlying postulates are false. Also they forget the starting points for these postulates.

A growing number of eminent physicists are objecting to the claims made by too many of the specialists in nuclear physics. The nuclear physicist Madame Tonnelat has written: " This engulfing of theory by formalism, and of physics by mathematics is nothing new ".

They have wanted to apply to biology laws of physics which obviously have never been considered in a biological context. A number of examples can easily be given.

Professor H. Prat, of the Faculty of Sciences has written: " In fact all our physical and biological laws have been more or less based on an implicit acceptance of the idea of identity. They all must be reconsidered and reformulated."

Statements by the physicist Brillouin could also be quoted. For example, relative to repair of tissues and reconstitution of parts of the body in a lower animal: " This is why so many scientists consider our present physical and chemical laws do not explain such strange phenomena."

The theories of Einstein have never been considered from a biological viewpoint. Even J.-M. Levy-Leblond, professor at the College of France, wrote: " The theory of General Relativity ... is characterised by having very few points of contact with concrete physics." Many physicists dispute the existence of even these few points which the Relativists claim as proving their law. These physicists consider that the results presumed to support the theory can very well be justified by other better established theories. R. H. Dicke in *The Theoretical Significance of Experimental Relativity*, New York, 1965, wrote: " At the present time there is disillusion with the infatuation which has arisen from these proofs." In France one hears of " disruptive aspects of physics " and of a " possible breakdown of Einstein's theory "; in England " a new look at the fundamentals of relativity may be called for," etc.

In atomic physics, dogmatic and puerile generalisations have been attempted in order to bring everything into Einstein's theory of Special

Relativity. This has led E. C. Zeeman, of Cambridge, to write: "a science often disintegrates in chaos because of an accumulation of new experimental evidence which contradicts existing theories. To give only one example, at this time nuclear physics is losing its way."

The more "reserved" scientists declare: "Einstein's equations remain valid, but it now appears that they are, perhaps, incomplete." (*The Sciences*, July 1964, published by the Academy of Science of New York.) Others make statements such as: "Subjection of our ideas to a fresh examination and a complete revision lead to the conclusion that there are adjustments to be made in the theory of Relativity."

But it should not be thought that I am against Einstein. I am only against the unreasoning disciples who have not understood their master and have wrongly applied the laws for which Einstein had set clearly defined limits. His laws are valid only for phenomena conforming with his initial postulates. In other contexts they only lead to absurdity.

Einstein was careful to insist on this aspect of his laws, but who now reads Einstein? One reads books on relativity, some written by professors who do not understand the subject, who mechanically apply calculations based on false premises.

Einstein himself said: "The validity of the law of Relativity should be accepted only when applied to certain specified bodies, Galilean bodies; those having a rectilinear uniform movement without rotation; and not to others having different movements. It is within this meaning that we speak of the law of special Relativity."

Of General Relativity, he said: "The frequently stated law of constant speed of light in vacuum, which is one of the two fundamental assumptions in the Theory of Relativity, is not of universal application.

"Its results are valid only when the influence of the fields of gravity on the phenomena can be ignored.

"The physical interpretation of the Euclidean spacetime continuum ... was made possible by the law of constant speed of light ... which could not be sustained by the theory of General Relativity."

In fact his law of Special Relativity was only applicable to a rectilinear and uniform movement.

One could continue to show that Einstein was well aware of the need to set limits to each of his theories. The law of Special Relativity, with its formula $E = mc^2$, is applicable only if the respective speed ("c" in the formula) of light or electro-magnetic waves is constant and equal to approximately 300,000 km per second and if no outside field modifies this speed, (he had only considered gravitation forces). But it is a serious error to seek to apply this law when external fields modify the speed of the wave. Such changes may occur for various reasons. In an ionised atmosphere the wave speed is about the same as that of light waves but only at very high frequency; at low frequency, the upper ionised atmosphere, the speed is around 50,000 km per second. The assumed speed, on which calculations for studying atomic nuclei are based, has not been ascertained to be the speed of light. The presence of protons also gives rise to a strong ionisation, and there are yet imperfectly known intense fields. Moreover, genuine physicists know that Einstein's laws were formulated for continuous magnitudes. There are difficulties in applying them to certain aspects of the atom where the particles are of varying magnitudes; quantum mechanics would here seem a more fruitful field of study. So the objections of certain physicists to my work do not merit attention. A large number of them recognise this and try to find a satisfactory physical explanation within the framework of our present knowledge.

"Satisfactory" explanations always come later. Meanwhile, it is enough to note that none of the current classical theories are entirely applicable to our work. The facts are there, indisputable, and no valid opposition exists.

It has been obvious for a long time that physics and chemistry of today are unable to explain life. Life is a struggle against certain physical laws such as entropy, or the breakdown of energy. Entropy brings cell degradation, levelling down of energy and death. It is recognised by many scientists, to cite them all is impossible. In *Vie, Matière et Information*, by Brillouin, published by Albin-Michel, Paris, 1959, the author writes: "The law of Carnot is essentially a sentence of death; it is ruthlessly applied in a world already going to die. Life, for a limited time, postpones this death sentence."

Gabriel Bertrand, whose name is associated with the study of trace elements and with whom I worked at the Conseil d'Hygiène in Paris for sixteen years, said, when I told him of my conclusions on transmutations: "You preach to the converted." He had already written in *Atomes*, July 1948: "We should accept that spontaneous generation, even of the simplest organisms, remains a mystery as difficult to explain by biochemistry as by microbiology. Does this mean we are confronted by an insurmountable barrier? I do not think so. Combining their efforts with those of the physicists, contemporary chemists have worked out certain transmutations of elements."

He realised, therefore, that biochemistry could not explain everything, and that transmutations must be taken into consideration. But he did not go into the subject very deeply; he was thinking only of the radioactive transmutations in nuclear physics. (Bertrand wrote in 1948, the first atomic bomb was in 1945.) However, it must not be said that I introduce a metaphysical concept of "vital energy". I have devoted my time and efforts to a demonstration of facts supported by biology. I appreciate that an investigation of this phenomenon (transmutations) in conjunction with physics has not yet reached a stage where a definite opinion can be formed. I cannot therefore take any fixed position regarding the type of energy involved in these transmutations. Many books are devoted to the study of biological energy, but the subject contains too many unknown factors to permit an essential understanding. It is not sufficient to say that this energy results from a chemical action of ATP (adenosine triphosphate).

The exact nature of the energy concerned in these transmutations is unknown. Friends in the physical sciences have attempted to find an adequate explanation—one within the confines of present day physics. I will not report their respective working hypotheses—which are varied—because, for the time being, I consider that a full and satisfactory picture of the phenomenon is not yet possible, either quantitatively or qualitatively. So far only outlines may be seen.

Einstein himself, before he died, had come to realise that his theory of General Relativity might not be of "general" application because he did not succeed in applying it to all phenomena. He declared that in his opinion, God did not play just one game but four or five. Einstein's theory is incompatible with electromagnetism and with certain aspects of quantum mechanics (without counting the fact that there are two different and opposing schools of Wave Mechanics). We have not yet been able to "build a bridge" between these theories which are not only divergent but, on several points, fundamentally opposed, though on others they are complementary. In 1964 de Broglie wrote that quantum mechanics "does not really give a reasonable explanation of certain essential experimental facts, and must be revised." (*La Theorie Dynamique de la Particule Isolée*, Gauthier-Villars, Paris.)

Before he died, Einstein was of the opinion that there were several quite distinct theories, each one applicable in its own special area. Today there are many passionate adherents of General Relativity, but also many who flatly oppose it. Physicists are divided, and I for one do not attempt to bring them together; the only important fact is that none of them has been able to prove to me that I am wrong. Even the law of Special Relativity cannot do this. No one has been able to show that the law applies in biology or that the basic assumptions utilised by Einstein are valid for biology neither does it apply fully inside the atom where there is something different.

It is generally accepted that in the atom there are both electromagnetic and electrostatic energies; but the electrically charged and displaceable particles (protons) move on their trajectory just like electric currents. In the diagram we have presented of the nucleus, the electrostatic activity of the particles is accompanied by an opposing electrodynamic activity. There are parallel currents in the same direction, attracting each other. This has been described in our previous books since these attracting electrodynamic forces are not mentioned in classic works on nuclear physics, which only speak of repelling forces between nuclear particles, in accordance with Coulomb's law. However, three years after I had exposed these forces, research scientists at the California Institute of Technology reported that these forces took part in weak interactions, evaluated as equal to 10^{-25}

116

times that of gravity in a nucleus and 10^{-14} times weaker than the binding force between protons and neutrons; but I have some reservations on the method of calculation adopted.

Inside the atom nucleus there are also " strong interactions "; these are the bonding energies between nucleons. These bonds, to which, in certain aspects, Einstein's theory of Special Relativity applies, have been the subject of much investigation. But again, though Einstein's theory may be valid for this particular case, it is a mistake to generalise by saying, for example, that matter can be transformed into energy. This statement is false, even though it is found in practically all text books on nuclear physics. We know only how to use the bonding energy between nucleons (which seems to come from the " mesons "). But matter is not transformed into energy; matter is essentially composed of protons and neutrons, and in atomic fission, the nucleons do not disappear, but are found intact in the fission products. If some neutrons are expelled they are not destroyed. For matter to disappear it must be opposed by anti-matter. To say so much energy is used up by the breakdown of so many kilograms of uranium is not true. First we would need to supply an equivalent energy for an anti-matter (or rather a matter-anti-matter coupling, since it is not possible to create a proton, for example, with energy). This means that creation or destruction of matter is a fiction, but one that is taught, unfortunately.

Besides, the word " matter " has no exact meaning; we just do not know what matter is; we do not know what a proton or an electron is made of; the word serves only to cloak our ignorance. Matter has not been proved to come from energy. The physicists come out with various hypotheses; L. Romani even questions if what are termed particles are not actually the gaps in the energy supply.

The " weak interactions " in the atom nucleus are only now being studied. Interactions occur in natural radioactivity but in some conditions they are still weaker (for example in disintegration of the K-mesons). They are imperfectly known, and are of a "capricious" nature, meaning that they cannot be explained by any former laws. We are ignorant of the forces giving rise to these energies and of the laws governing weak interactions. (The physicist R. de Puymorin considers that they result from known forms of energy and should not be separately studied; they could in some way be forces of slightly different amplitudes in opposite phases.)

Inside the atom nuclei there may, perhaps, be other energies of a totally unexpected nature. For the time being, the analytical investigation of the nucleus has reached a dead end; a full consideration of all relevant factors is required. Be that as it may, if the low energy transmutations we have demonstrated can be explained by a coherent theory of current physics, such a theory must, in my view, be sought in the field of weak interactions. And this would be quite inconsistent with classical nuclear physics with its hypotheses based on powerful interactions.

With weak interactions, the established laws are completely upset; there is no assurance of conservation of energy, the mass/energy equivalent does not exist; and it is possible to have a simultaneous variation of mass and energy in the same direction (see *Low Energy Transmutations*). Even with strong interaction there can be a simultaneous increase in both mass as well as energy. This has been demonstrated by J. Debiesse, director of the Centre for Nuclear Studies at Saclay, with regard to the exchange of pi-mesons (π or pions) between protons and neutrons: "Apparently these mesons are emitted by a rapid increase of mass and energy in the system." Brillouin also has written: "No one denies that conservation of energy is an experimentally proved fact, but no one gives us the right to attribute more generality and more importance to this principle than to the experiments which demonstrate it."

It was recognised in 1964 that weak interactions violate the law of time symmetry. This law states that the force between two moving particles is the same if their speed is reversed, providing that the direction of their rotation is also reversed. This law is valid for gravitation, for electro-magnetism, and for the powerful reactions (the so-called nuclear forces); but in weak interactions we are concerned with a quite different and unknown form of energy; remember also that the law of parity is violated.

Thus, little by little, the more eminent atomic physicists[1] are making public statements to the effect that in the atomic nucleus there are unknown factors not in accordance with theories established according to the earlier state of our knowledge. The departure from the law of time symmetry in the breakdown of meson K°_2 would involve an energy of only 10^{-8} eV; could not this be regarded as an approach for the study of negative entropy of life? It would be a confirmation of what Bergson wrote: "All analyses show that life is an effort to reclimb the slope which matter descends." Only the pretentious and those tied to childish dogma can say they know everything, and, in the name of "well-known laws", which they generalise, forgetting the hypotheses on which these laws were based, reject as impossible biological transmutations and all other low energy transmutations. But there they are on territory they have not explored, and of which they know nothing. "This explanation (of vital phenomena as interpreted by physico-chemical laws) is accepted as dogma by a certain materialistic school." Brillouin.

The genuine scientist who does not know my work makes no pronouncement on it. One should only judge what one knows well; the reading of one article in a review is not enough. A true scientist will never contest any aspect of my work unless he has himself repeated some of the experiments I have described. However, no one has been able to prove me wrong.

Those who deny the facts without testing them, should not interest

[1] Those of good faith, for not all of them are. Science has become a trade and no longer a vocation; scientists are in no way "supermen".

us; they will always deny, whatever the evidence, since that is their nature, but it is not they who will read this book. May those who doubted through lack of precise facts find in the preceding pages sufficient material to convince them or, if not, to incite them to make their own investigations before coming to a verdict.

An open-minded scientist is modest and will acknowledge his ignorance. In the field we have touched upon there is yet much to explore. It is a field worthy of close investigation and constitutes a great opportunity for young research workers. One thing is sure, here are experimental results; if they cannot be explained and fitted into the " accepted " laws, so much the worse. In that case we will wait for the laws to be changed, and, meantime, make use of the rules already admitted.

Those who believe that facts can be rejected if they do not conform to an "established law", should remember the intellectual courage shown by the great English astronomical physicist Hoyle, famous for his theory of the stationary universe. He has now rejected the theory which made him famous, (*Nature*, October 9th, 1965): "Recent Developments in Cosmology"). By a process of self-criticism, he found that the results which led to the discovery of "quasars" no longer permit acceptance of the hypothesis he utilised twenty years previously. (See also *Sciences*, December 1965.)

Hoyle states: "Recent progress has reinforced this conviction and the general opinion tends to agree that the equations in physics have a common singularity "; but he will not generalise. On the other hand, I see no objection to the assumption that present laws of physics are incomplete, which they almost certainly are. The problem therefore is to modify these physical laws in order to prevent this universal singularity or, in other words, to prevent the collapse of physics."

It is not only classical physics that is in danger. Hoyle recognises that if future observations confirm that physics has taken a wrong direction, " The properties of matter, the laws of chemistry, for example, would be completely changed."

SUPPLEMENT TO THE ENGLISH EDITION

We give, as an additional example, the summary of an experiment that the author conducted (after the publication of the French edition) to ascertain whether the increase of an element in an animal is the result of its extraction from its environment: [1]

NON-ZERO BALANCE OF PHOSPHORUS AND COPPER IN THE LOBSTER AFTER MOULTING

REPORT ON AN EXPERIMENT

Lobsters (*Homarus vulgaris*, or *H. gammarus*) weighing about 200g, in the inter-moult period, are placed in tanks of normal sea-water until they have finished moulting. Lobsters of this size, moult about twice a year. Just after moulting, one of them was removed from the water, wiped, weighed and then incinerated in an electric oven at 500°C. The ashes were treated with HCl (17.5 pure HCl in 500 cm³) and the analyses, for different proportions, were carried out in this solution.

Another lobster, as soon as moulting was finished, was weighed and then placed in a tank of sea-water that had not contained another animal: its normal content of phosphorus and copper was therefore complete.

This water originally contained 1.27 g/litre of magnesium which was reinforced by the addition of ⅔ magnesium chloride and ⅓ magnesium sulphate, so that the Mg content rose to 2.71 g/1 (average result of two methods: by black eriochrome complexometry and by yellow thiazol colorimetry). This yield is only an indication, and can largely vary. The calcium of this water had been precipitated by chloranilic acid, to give 50 mg/1 instead of 430 mg/1 in the unpurified sea-water, and the pH was adjusted with soda to 7.35 to start with.

The lobster was left for seventeen days in this water without food, but we wished to avoid having to take into account the effect of the mineral contribution by food. The tank contained 75 litres of water, including filter. This water was in a closed circuit; a siphon causes it to circulate permanently over an activated charcoal filter coated with glass silk. The filter is designed to hold back the solid excreta and gases of the lobster's metabolism. The water is aerated by a rotary electric pump which blows air under pressure to the bottom of the tank to re-oxygenate the water. Seventeen days after

[1] See also: The Journal of the Soil Association—January 1970

moulting, the lobster was removed from the water, weighed, incinerated and analysed as above. Our initial analysis had shown that neither the filter nor the pipes (plastic) could elute either phosphorus or copper (by ionic exchange).

RESULTS OF THE ANALYSIS

A—PHOSPHORUS

1 BALANCE

The sea-water contained 0.0020 ± 0.0001 mg/litre of P, a relatively low figure; it was taken near the coast of South Brittany. This rate gives a total weight of P, in the 75 litres of the tank, of 0.150 mg.

	At time of moulting		17 days afterwards	
In the lobster	380	mg	430	mg
In the water (751)	0.150		0.150	
	380.150	mg	430.150	mg

In aliquots, the phosphorus was found by three methods: ceruleo-molybdic colorimetry, ascorbic acid + sulfomolybdic acid, and by phosphomolybdic acid + benzidin acetate. The analyses of water, and/or of the animals, were all carried out three times by these different methods, and the figures quoted are averages.

The weight of P at time of moulting is obtained by a proportional rule checked by two sets of weighings: weight of the animal in the inter-moulting period, at the moment when the experiment starts (animals of the same sex and about the same weight were selected—in this case males). The animals were weighed just after moulting (loss of the exuvium); the difference between this weight and that at the inter-moulting time, for these creatures about the same size, weight and age was found to be almost constant, with sometimes only 1% deviation. Thus it was possible, by a proportional rule, to equate the weights to a constant weight after comparison between a test animal killed as soon as its moulting finished, and an animal placed in water with reinforced magnesium for seventeen days.

2 COMMENTS

In seventeen days the lobster gained 50 mg of P (in another experiment, a similar lobster, a little smaller, gained 30 mg in thirteen days). That is an increase of 13% in seventeen days. Now, in all the water of the tank there was only 0.150 mg of phosphorus in seventeen days. It therefore incorporated a quantity of phosphorus more than three hundred times higher than that contained in the environment. An endogenous formation of phosphorus, by some unknown enzymatic process, seems the only possible explanation (it should be mentioned that we observed a similar result in mice which had been given an excess of magnesium in their diet; see: *Comptes-Rendus de l'Académie d'Agriculture de France*—13.12.1967).

B—COPPER

1--BALANCE

The copper content of the raw water was 0.026 mg/litre (1.95 mg/ 75 litres). At the end of the experiment (seventeen days after the moulting) it was found to be 0.066 mg/1 (4.95 mg/751).

	At the time of moulting	17 days afterwards
In the lobster	3.40 mg	5.51 mg
In the water (751)	1.95	4.95
	5.35 mg	10.46 mg

Increase (total) $\sim$ 95% (62% in the lobster; 131% in the water).

Copper found on aliquots: by 2-2^1 diquinolyl colorimetry; and by addition of cuprethol in a tray containing 10 ml at 434 m μ (without extraction); but the figures quoted above were obtained by readings at 25°C in a Beckman spectrophotometer with atomic absorption; they were very near the average figures obtained by chemical methods: we examined by means of a Beckman unit in order to check the chemical analysis by a physical method, and to prove that chemists do not make mistakes in their counts.

2 COMMENTS

Copper is an important element in crustaceans (and other marine creatures) since their blood (haemolymph) contains haemocyanin, an oxygen exchange protein with copper, instead of iron in haemoglobin. Already, various investigations have been carried out on these animals, but it was implicitly accepted, without experimenting, that the copper came from the environment. Our experiments show that this is not the case, because the lobster contained 2.11 mg more Cu whereas there was only 1.95 mg in all the tank water and this could not account for such an increase of Cu in the animal.

We must add that previous research had shown a link between copper and iron. Therefore we analysed the iron in the water at the beginning of the experiment and then at the end (seventeen days after). Analysis by chemical methods was confirmed by spectrophotometry with atomic absorption. We found respectively 0.0845 mg/1 and 0.069 mg/1; or, in 75 litres 6.337 mg and 5.175 mg; that is a decrease of 18% (in the raw water Fe/Cu = 3.25).

CONCLUSION *

It seemed to us useful to go fully into the results of this experiment which had not previously been carried out, even in principle. We hope that others may repeat a similar experiment on a larger number of animals in order to confirm or refute the results. After the first fortnight following the first moulting in the tank it can be seen that the total content of P and Cu in the water is insufficient to explain the increase of these elements in the animal which had certainly not taken them only from the medium in which it lived. It is therefore likely that an endogenous production of these elements takes place —inexplicable so far by the laws of physics and chemistry included in the present teaching curriculum. The explanation can only be a biological transmutation, by an enzymatic action whose process remains to be defined; but how much more of the unknown is there in the sphere of enzymes?

---

* We do not list here the Ca balance (see *Rev. de Pathologie comparée et de Médecine Experimentale* Sept/Oct 1969)

ACKNOWLEDGEMENTS

We have pleasure in thanking Mr. Gwenaël Bolloré for his generous help during these investigations, for the breeding and examination of the lobsters at his Oceanographic Museum at Odet, near Quimper (Brittany), France.

SHORT BIBLIOGRAPHY (English texts only)

Adley, F. E., *J.ind.Hyg.Toxicol.* Jan. 1946.

Baranger, P., " Do plants effect the transmutation of elements?" *J. the Soil Assn.*, April 1960.

Baranger, P., " Variations in the phosphorus and calcium content in vetches during germination ". *J. Biol.* Dec. 1960, Bombay.

Branfield, W., *Continuous Creation.* Routledge & Kegan Paul. London, 1950.

Butler, J. A., " Life and the Second Law of thermodynamics." *Nature,* Aug. 1946—158—153, 154.

Daugatry et al. *Symposium on the geochemistry of sedimentary carboniferous rocks.* Copenhagen. 1960.

(in same book, on the same subject, S. Seibold).

Erlich, H. L., " Microbial participation in the genesis of manganese nodules." Rensselaer Polytech. Instit. Troyes, N.Y. 1961.

Erlich, " Bacterial action on manganese in nodules enrichments ". *Appl. Microbiol.* 1963 II,1.

Grebe, J. J., " A periodical table for fundamental particles." *Ann. N.Y. Acad. Sci.* 76, 1. Sept. 1958.

Grebe, J. J., " Easy as Pie ", *Chem. Engng News,* May 5, 1958.

Grebe, J. J., " Power from Nuclear Reactors ". *Electl Engng, N.Y.* Sept. 1961.

Hauschka, Rudolf. *The Nature of Substance.* Vincent Stuart. London. 1964.

Heighton, H. H., " An Electrical model of Matter." E. I. du Pont de Nemours. Jan. 26, 1959.

Kervran, C. L., " Increase in phosphorus and copper in the lobster after moulting ". *J. the Soil Assn.* Jan. 1970.

Kervran, C. L., " About the above experiment ", *New Scientist.* Aug. 13, 1970, p. 359.

Kruse et al. " Study on magnesium deficiencies in animals ". Symptomatology resulting from magnesium deprivation. *J. biol. Chem.* 96 529, 1932.

McCollum and Orent. " Effects on the rate of deprivation of magnesium ". *J.biol.Chem.* 92 XXX, 1934.

Maxwell. W., " On the behaviour of fatty bodies and the role of lecithines during normal germination ", *Am. chem. J.* 13, 16, 24. 1891.

Mero, J. L., " Mineral Resources of the sea ". *Trans. N.Y. Acad. Sci.* 26, 525-544. 1964.

Miller, Er., " A physiological study of the germination of *Helianthus Annuus.*" *Ann. Bot.* 24: 693, 726. 1910.

Odagiri, M., *Chemical Superposition,* State Laboratory of Theoretical Chemistry, Kinki University, Osaka, Japan. 1961.

Piccardy, G., *The Chemical Basis of medical climatology*. Thomas, Springfield. 1962.

Prout. "Some experiments on the change which takes place in the fixed principle of the egg during incubation." *Phil. Trans. R. Soc.* June 1822.

Puymorin, R. de, "The electronic origination of gravity". *Bull. U.I.R.T.* Oct. 1963.

Rosenauer, H., "What are biological transmutations?" *J. the Soil Assn.* Oct. 1966.

Sabetay, H., "Biological Transmutations". *Sc. Group J.* London. Jan. 1964.

Sabetay, H., "A recent scientific discovery: Natural Transmutation". *Canadian Theosophist*, Toronto. May 1965.

Schrödinger, *What is life?* Cambridge University Press. 1944.

Selye, H., *Calciphylaxis*. Chicago. 1962.

Szent Gyorgÿi, *Introduction to a submolecular biology*, 1960.

Turkee et al., "Chemical Transformation of phosphorus in the growing corn plant". *Agric. Res., Wash.* O 46, 12. 1933.

INDEX

Bones, fossilised 76
  fractures 74
  mineral salts content 76
Boron 5, 29, 102
Branfield, W.: *Continuous creation*
  vi, ix, 15
Brewers yeast 42, 45
Brillouin, physicist 113
  on conservation of energy 118
  Vie, Matière et Information 115
Bromine 5, 69
Bronsted, on acids 9
Bulletin of I.B.I.N. 88n.

Cailleux, André: rock research 27
Calcium v, viii, ix, xii, xiii, 1, 5, 15,
  25, 26, 27, 28, 31, 34, 37, 46,
  50, 55, 59, 60, 63, 64-65, 74, 75,
  77-78, 83-84, 92, 106
Calcium carbonate 88
Calcium fertilisers 89
Calcium metabolism 78
Calcium salt 68
Calcium sulphate 33
California Institute of Technology
  116
Calories, values 53, 75
Cambodia, sandstone monuments
  48
Carbon 5, 8, 28, 29-30, 37, 48, 90,
  95
Carbon dioxide 8
Carbon monoxide x, 15-23, 39, 40
Carbonates 32
Carbonyl grouping 40n.
Cardan, Jerome 72
Caries, dental 76
Carnalite 58
Carnot-Clausius, on breakdown of
  energy 13-14
Cave microfauna 91-92
Cell, degradation 115
Cement, natural (Portland) 72-73
Centre National de la Recherche
  Scientifique 91
Chara foetida 10
Charnot, Dr 47, 64, 76, 77
Charruyer, Dr, on calcite geodes 73

*The chemical basis of medical
  climatology* 10
Chemistry 1, 113
  biological vii, viii
  and living organisms 2-4
  symbols 4-5
Chlorellae 69
Chlorine 5, 38, 69
Chlorophyll 11n., 60, 84
Choubert, G. 27, 69n.
Clarkson, T. W. 78
Clater xiii
Clay 29, 91-92
  'Clusters' 39
Coal deposits 29, 30, 95
Cobalt 13
Colloids 11, 12
Combinations, chemical 6-9
Compost, source of potash 82
Conseil d'Hygiène de Paris 21
Continuous creation, W. Branfield
  vi, ix, 15
Copenhagen Symposium, 1960, on
  dolomite rocks 73
Copper 2, 5, 71, 76, 89, 120-122
Coral 32
'Core' of nucleons 101-102
Coulomb's law 116
'Coupling' of nuclei 37-40
Crabs, formation of shells 61
Crop rotation 24
Crops v, 90
Crustacea 61, 122
Cultural media, experiments 42-46
Cussonneau, E., analysis of soils 88-
  89
Cybernetics 14
Czapeck 15, 43

Dahiez, analysis of plants given
  sulphur 89
Daisies, in lawns 25
Daugatry 73
de Barjac, nitrogen research xii
de Broglie, Louis viii, 14, 116
de Cayeux, A. ix, 32, 69, 72-73
de Larebeyrette, J. 11, 12
de Puymorin, R. 117